GRASSLANDS OF THE MONSOON

by the same author

published by Faber & Faber

R. O. Whyte
CROP PRODUCTION AND ENVIRONMENT

R. O. Whyte
MILK PRODUCTION IN DEVELOPING COUNTRIES

R. O. Whyte and M. L. Yeo
GREEN CROP DRYING

other publishers

R. O. Whyte
LAND, LIVESTOCK AND HUMAN NUTRITION IN INDIA
(*Praeger*)

R. O. Whyte
THE GRASSLAND AND FODDER RESOURCES OF INDIA
(*Indian Council of Agricultural Research*)

R. O. Whyte, G. Nilsson-Leissner and H. C. Trumble
LEGUMES IN AGRICULTURE
(*FAO Agricultural Studies*)

R. O. Whyte, T. R. G. Moir and J. P. Cooper
GRASSES IN AGRICULTURE
(*FAO Agricultural Studies*)

R. O. Whyte and G. Julén
PLANT EXPLORATION AND INTRODUCTION
(*Genetica Agraria*)

R. O. Whyte and M. L. Mathur
THE PLANNING OF MILK PRODUCTION IN INDIA
(*Orient Longmans, New Delhi*)

GRASSLANDS OF THE MONSOON

R. O. WHYTE

FABER AND FABER
24 Russell Square
London

First published in 1968
by Faber and Faber Limited
24 Russell Square London WC1
Printed in Great Britain by
Latimer Trend & Co Ltd Plymouth

S.B.N. 571 08583 0

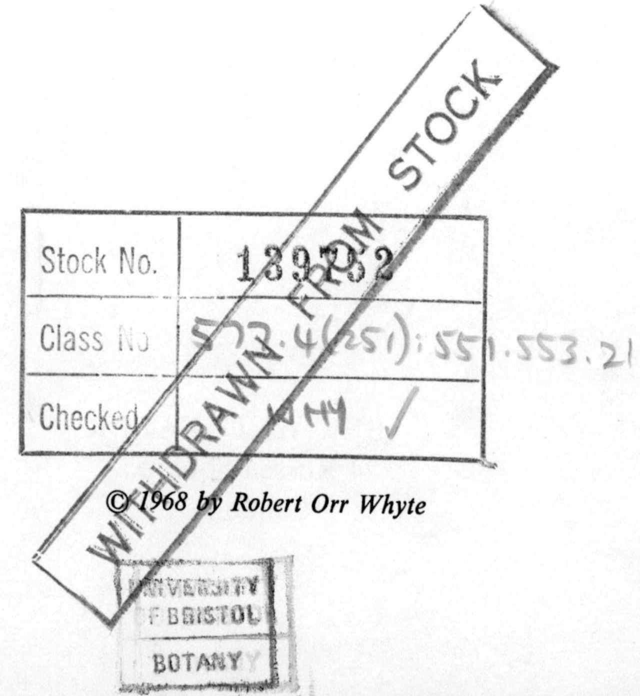

CONTENTS

IV–XIII *The Transect*	*Zones of Contrast*	
VI. Southern Asia		*page* 166
India		166
West Pakistan		186
	Tibet	193
Nepal		198
Burma		202
Ceylon		205
VII. China (Mainland)	China	213
VIII. Far East U.S.S.R.		218
IX. China (Taiwan)		219
X. Japan		223
XI. Malesia		224
XII. Papua and New Guinea		244
XIII. Australia	Australia	252

ILLUSTRATIONS

PLATES

FIGURES

TABLES

ACKNOWLEDGEMENTS

The author gratefully acknowledges his thanks to the following for permission to reproduce maps:

Figures 2 and 3, from the FAO African Survey *Report on the Possibilities of African Rural Development in Relation to Economic and Social Growth*, Rome, Food and Agriculture Organization of the United Nations, 1962. Also reproduced in *Arid Lands*, Methuen/UNESCO, 1966.
Figures 5 and 16, from E. H. G. Dobby, *Monsoon Asia*, Quadrangle Books Inc. and University of London Press Limited.
Figure 4, from G. B. Cressey, *Asia's Lands and Peoples*, Third Edition 1965. McGraw Hill Inc.
Figure 6, from Pierre Dansereau, *Biogeography and Ecological Perspective*, Ronald Press, New York.
Figures 8, 9 and 10, from J. Beaujeu-Garnier, *Trois Milliards d'Hommes*, Librairie Hachette.
Figure 12, from P. Pédelaborde, *Les Moussons*, Librairie Armand Colin.
Figure 18, *The Times*, London.
Figure 35, from Keith Buchanan, *The Chinese People and the Chinese Earth*, G. Bell and Sons.
Figures 40 to 50 inclusive, from C. G. G. J. van Steenis (papers in *Bull. Jard. Bot. Btzg.* and *Reinwardtia*).
Figure 53, from *The Australian Environment*, Third Edition, C.S.I.R.O./Melbourne University Press.

CONTRIBUTORS

Miss E. J. Abbott, Pasture and Fodder Crops Branch, Plant Production and Protection Division, FAO, Rome, Italy.

M. Batisse, Director, Division of Scientific Research on Natural Resources, Department of Natural Sciences, UNESCO, Paris, VIIe, France.

Dr. Erna Bennett, Scottish Plant Breeding Station, Pentlandfield, Roslin, Midlothian, Scotland.

S. T. Blake, Botanic Garden, Brisbane, Queensland, Australia.

A. W. Bogdan, Commonwealth Bureau of Pastures and Field Crops, Hurley, nr. Maidenhead, England.

Dr. N. L. Bor, c/o Royal Botanic Gardens, Kew, Richmond, Surrey, England.

J. P. M. Brenan, Deputy Director of Herbarium and Library, Royal Botanic Gardens, Kew, Richmond, Surrey, England.

Dr. R. K. Brummitt, Royal Botanic Gardens, Kew, Richmond, Surrey, England.

Professor Keith Buchanan, Victoria University of Wellington, Geography Department, P.O. Box 196, Wellington, New Zealand.

W. D. Clayton, c/o Royal Botanic Gardens, Kew, Richmond, Surrey, England.

Professor Monica M. Cole, Bedford College, Regent's Park, London N.W.1, England.

P. M. Dabadghao, Indian Grassland and Fodder Research Institute, Jhansi, Uttar Pradesh, India.

Miss I. R. Dudeney, Division of Scientific Research on Natural Resources, Department of Natural Sciences, UNESCO, Paris, VIIe, France.

Dr. Alexander Gilli, Penzingerstrasse 56, Vienna 19, Austria.

F. N. Hepper, Royal Botanic Gardens, Kew, Richmond, Surrey, England.

Dr. C. E. Hubbard, formerly Keeper, Herbarium, Royal Botanic Gardens, Kew, Richmond, Surrey, England.

A. Johnston, Research Station, Lethbridge, Alberta, Canada.

Professor M. Kassas, Faculty of Science, Fouad I University, Cairo, U.A.R.

M. D. Kernick, c/o United Nations Development Program, P.O. Box 1555, Teheran, Iran.

Professor H. G. Kmoch, Institut für Pflanzenbau, University of Bonn, Katzenburgweg 5, Germany.

Librarian, Commonwealth Scientific and Industrial Research Organization, P.O. Box 109, Canberra, ACT, Australia.

Dr. L. Mattsson, Director, UNESCO South-east Asia Science Cooperation Office, Djakarta, Indonesia.

Professor Th. Monod, Muséum national d'Histoire naturelle, 57, rue Cuvier, Paris 5, France.

Ch. Monod de Froideville, Rijksherbarium, Schelpenkade 6, Leiden, Netherlands.

M. Numata, Department of Biology, Chiba University, Japan.

H. Pabot, formerly c/o United Nations Development Program, P.O. Box 1555, Teheran, Iran.

Dr. R. A. Perry, Division of Land Research, CSIRO, P.O. Box 109, Canberra, ACT, Australia.

Dr. R. A. Peterson, Chief, Pasture and Fodder Crops Branch, Plant Production and Protection Division, FAO, Rome, Italy.

Professor R. Portères, Department of Ethnobotany, 57, rue Cuvier, Paris 5, France.

Mahendra Prakash, Office of Conservator of Forests, Working Plan Circle, Jaipur, Rajasthan, India.

Professor P. Quezel, Faculté des Sciences de Saint-Jérome, Traverse de la Barasse, Marseille 13, France.

Professor K. H. Rechinger, Naturhistorisches Museum, Burgring 7, Vienna 1, Austria.

Ch. Rossetti, 135 rue St. Dominique, Paris VIIe.

CONTRIBUTORS

P. Roy, Institut français de Pondichéry, India.

Y. Satyanarayan, FAO Ecologist, c/o Plant Production and Protection Division, FAO, Rome, Italy.

S. K. Seth, Forest Department, Uttar Pradesh, Lucknow, U.P., India.

C. L. Skidmore, Director, Commonwealth Bureau of Pastures and Field Crops, Hurley, Maidenhead, Berkshire, England.

L. P. Smith, Meteorological Office, London Road, Bracknell, Berkshire, England.

C. Tardits, 7, rue F. Mouthon, Paris XVe, France.

Dr. Tsing Liu, Division of Forest Biology, Taiwan Forestry Research Institute, Botanical Garden, Po Ai Road, Taipei, Taiwan.

Dr. C. G. G. J. van Steenis, Director, Rijksherbarium, Schelpenkade 6, Leiden, Netherlands.

C. W. Wang, Associate Professor, Forest Genetics, University of Idaho, Moscow, Idaho, U.S.A.

Chapter I

THE NATURE AND SIGNIFICANCE OF THE TRANSECT

From Dakar in Senegal to the Darling Downs in Queensland there extends a belt of natural or disturbed vegetation incorporating various types of grass covers. These are sufficiently similar to each other and sufficiently different from adjacent areas to the north and/or south to justify an attempt to consider whether they do actually represent a plant geographical zone or ecological transect. Allowing for variations due to rainfall, the types of grass cover in this transect or cross-section through Africa and Asia into Australia are generally similar in their botanical composition and ecological status and behaviour, and also to some extent in their tree and shrub associates.

The transect begins in West Africa, where the classic zones of ecoclimates (habitats with comparable climates) and vegetation extend from the Atlantic roughly to the Nile (Fig. 2). It continues in north-east Africa through the Sudan, Egypt, Ethiopia and the countries of the Horn of Africa. In other parts of Africa south of the Sahara, partly down the eastern side, we find similar species and communities which will be the subject of further discussion. In the winter-rainfall zone north of the Sahara along the Mediterranean coast, the genera, species and communities are completely different from those of the transect in the Saharan, Sahelian and Sudanian zones to the south.

The belt continues to the east, reappearing in Israel (see vegetation map Fig. 22) and where the tropical monsoonal influences extend from Bab el Mandeb up the Red Sea coast and some way into Arabia, along the Hadhramaut, and into the Arabian or Persian Gulf as far as Kuwait on one coast and the Makran on the other.

Then comes the geographical expansion in monsoon Asia (Figs. 4, 5), in Pakistan, India, Ceylon, Nepal and Burma. This vast zone of monsoonal grass covers in a forest climax is

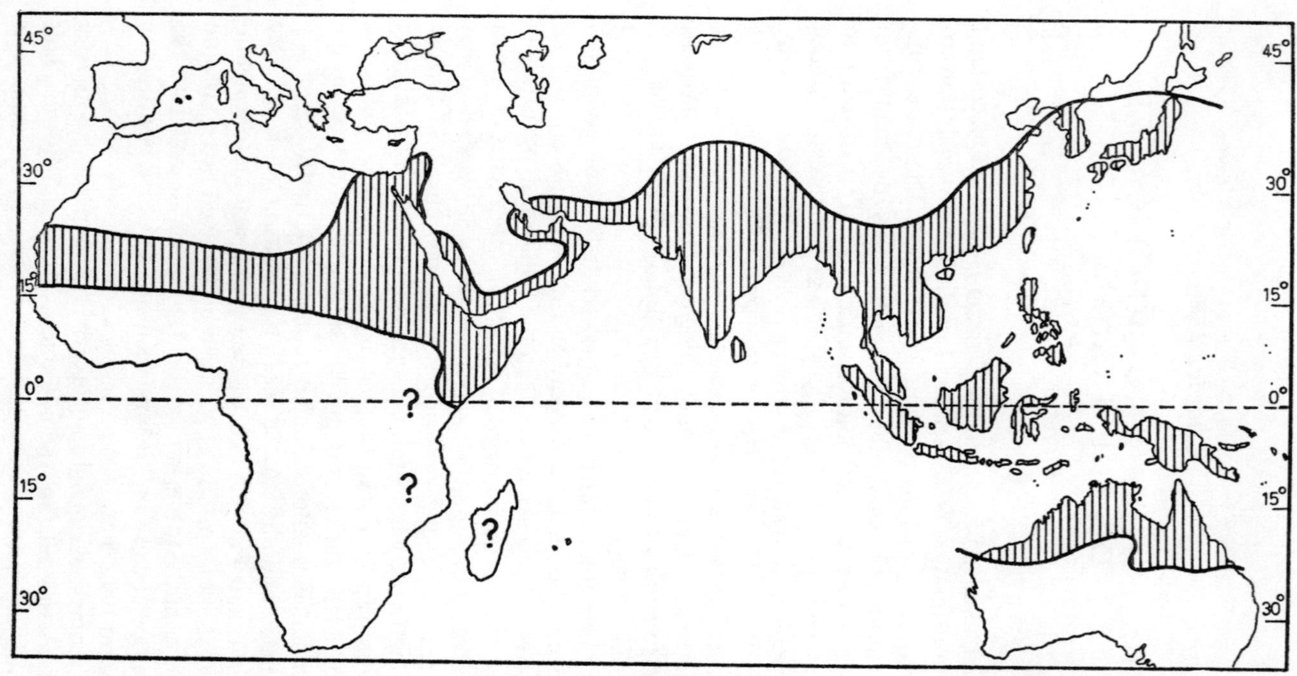

FIGURE 1
The transect of monsoonal grasslands
(see also FIGS. 14, 21 and 55)

bounded on the west by the Near Eastern Irano-Turanian vegetation of Iran and Afghanistan. The border runs up from the Arabian Sea coast through what was Baluchistan, to continue as a catena-like formation in the border zones between the North-west Frontier Province of West Pakistan and Afghanistan (Fig. 33). The border then becomes lost in a mosaic of areas with Mediterranean, Irano-Turanian or monsoonal climates in the western Himalaya.

Throughout the length of the Himalaya from west to east, the monsoonal grass species extend a certain distance up the southern slopes, to about 3,500 m., depending on exposure, distance from the plains of India and other factors. At about this elevation they meet and intermingle with the genera and species of an entirely different vegetation lapping over from the north, the hardy high-altitude species of the plateaux of Tibet (Sitsang) and the grassland west of China—the heartland of Asia.

The same or different monsoonal species are found again in the countries of mainland south-east Asia (Burma, Laos, Cambodia, Thailand, Malaysia) and in the south-east Asian islands of the botanical province of Malesia. They continue east into China—certainly into the Upper Burma/West Yunnan province recognized by H. Handel-Mazzetti,[79] and probably in all parts of Mainland China south of the Nanling which are regarded as an eastward extension of the Himalaya, or even as far north as the Tsinling (Figs. 4 and 5 show the limits according to G. B. Cressey[56] and E. H. G. Dobby[65]), and into Manchuria and the Far Eastern provinces of U.S.S.R. (see Chapter VIII).

This vast zone of comparable species and communities continues to a lesser extent in Taiwan and Japan, occurs spasmodically through the mosaic of seasonal and ever-wet climates of the equatorial archipelago of Malesia including Papua and New Guinea, and has a final full development in Western Australia, Northern Territory and Queensland in the monsoonal north of the Australian continent.

It may be asked: what happens in the Americas? It appears that the transect is unlikely to occur there to any extent because the geographical conditions do not promote the occurrence of true or false monsoonal conditions. On the one hand, the land mass is not so great as in Asia; on the other, most of the high

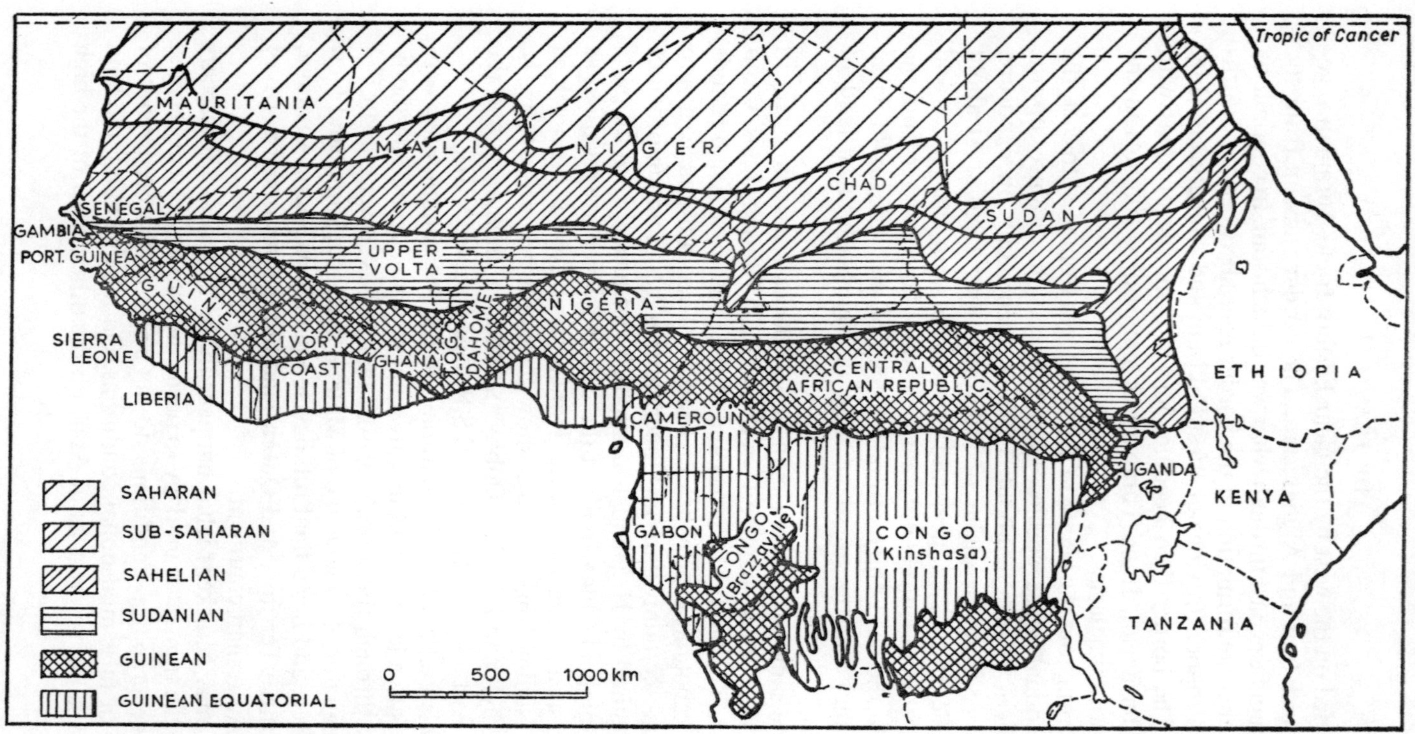

FIGURE 2
Ecoclimatic gradient in the western and western equatorial region of Africa

land systems run north and south, and do not therefore have such a marked effect upon pressure systems and the seasonal alternations in climate which characterize a monsoonal environment. It is to be hoped that a plant geographer in North or South America will review the situation in these countries and ecoclimatic regions.

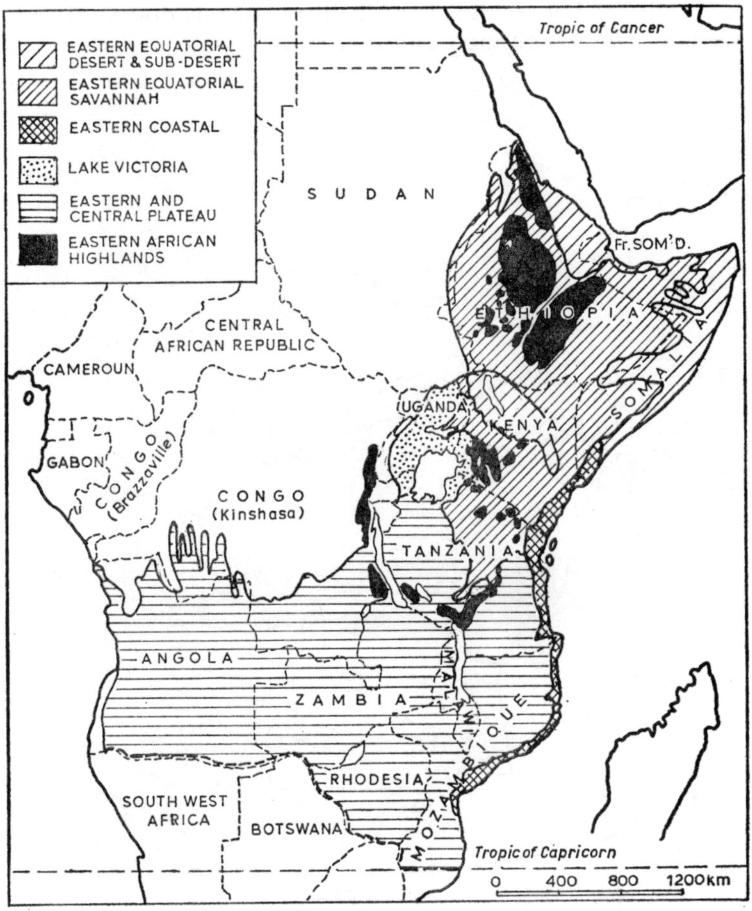

FIGURE 3

Ecoclimatic gradient in the eastern and central region of Africa

Terminology

An apology must be made for inexactitude in the title of this book, which had to be made intelligible to the general reader, with a minimum of ecological jargon. Grassland is a term applied to a particular type of vegetation in relation to its physiognomy. By far the greatest part of the grasslands discussed here would be more accurately called types of grass covers. A mon-

FIGURE 4
Monsoon Asia, according to G. B. Cressey

soon is a wind, which may bring wet or dry conditions. The use of the term in the title is not intended to indicate that our grasslands are airborne. The ambiguity which exists with regard to the term *monsoon* is discussed in the next chapter.

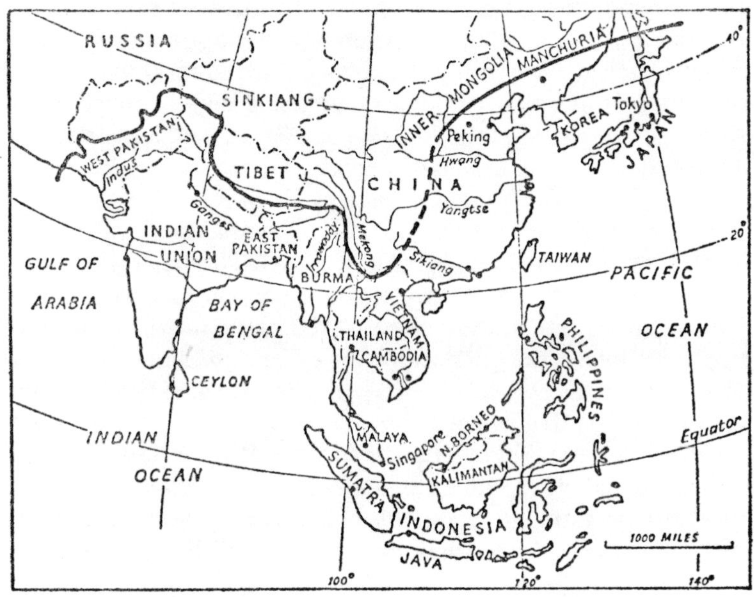

FIGURE 5
Monsoon Asia, according to E. H. G. Dobby

Global Ecological Zones

There has in recent decades been a somewhat haphazard tendency to consider regional and global problems on some kind of ecological basis. We talk about humid temperate Western Europe, the Mediterranean region, wet and dry tropics, monsoon Asia, and so on. The UNESCO approach comes nearer to an ecological definition of the environment than most, with its arid and semi-arid zones, and humid tropics. The continental and regional approach of FAO is political or ethnological, certainly not ecological—it is largely accidental that the Asian and Far Eastern region coincides with the limits of monsoon Asia.

It is the object of this book to propose that there is a vast zone of the world with a monsoonal climate, vegetation and characteristic forms of land use, cropping and animal husbandry. If we can as a result define this zone more accurately by association between specialists in ecology, agroclimatology and meteorology, it would be possible to proceed to the definition of other successive zones in which agroclimate, vegetation and land use systems are sufficiently uniform to give them some kind of unity. Studies of this kind would lead to the grouping together of specialists who could talk the same ecological language and who are faced with comparable problems.

It will be seen in reading Chapters IV to XIII that there has been great variation in the approach of the specialists who have worked at different sections of the transect. This has varied from studies of the floristics, phytogeography, physiognomy, ecological succession and the effect of the biotic factor, to speciation, and intercontinental and inter-island migration of plants through geological ages or following the movements of men by land or by sea. It has been the objective of the writer to bring together an assembly of disparate information, primarily from published and other sources that are not widely available, rather to stimulate thought, argument and further research, than to present a coherent and finalized statement of the situation.

Definition of Botanical Terms

Plant geography, geobotany, vegetation types, floras or floristics, phytosociology, bioecology—anyone not specialized in the overall subjects to which these terms refer may be excused at feeling somewhat confused. And into these fields of pure botanical science come the practitioners, the foresters with their forest types, and now the grassland ecologists expressing their heterodox approach by inventing the new concept of types of grass cover.

R. Good[76] has clarified the situation with regard to the scope and correct usage of some of these terms, as follows:

> Within the science of botany, plant geography is most intimately connected with plant ecology, these together making up the wider subject of *geo-botany*, which comprehends all aspects of the relation between plants and the surface of the

earth that is the substratum of their lives. Plant ecology is particularly concerned with the way in which plants are mutually related to one another and to the conditions of their habitat. Plant geography, on the other hand, is concerned primarily with the correlation between plants and the distribution of external conditions. The former is essentially physiological; the latter is essentially geographical. Expressed in another way, the difference is that between *vegetation* and *flora*, and a clear understanding of these two terms is important.

The chief features of vegetation reside in its quantitative structural characters because of their obvious influence on all other kinds of associated life. These structural characters are closely related to climatic conditions, and hence the same kind of vegetation, that is to say the same kind of dominant growth form, tends to recur in many parts of the world. For example, deciduous woodland is found not only in the British Isles and other parts of Eurasia, but also in parts of North America, as well as elsewhere, and in all these places it possesses much the same general features and dimensions. An attempt to show the distribution of *vegetation* types in simple form is made in Plate 2 (of Good's 'The Geography of Flowering Plants'), but it must be remembered that maps of this kind are seldom wholly satisfactory because of the ever-increasing difficulty of distinguishing between natural vegetation states and those induced by human activities.

The word *flora* is a purely scientific term and therefore has no common usage (which is itself an interesting commentary on the conception behind the word *vegetation*), and its meaning is best expressed by extending the example employed in the last paragraph. Although the deciduous woodlands mentioned there are alike in their vegetational features they will be found on closer examination to differ greatly and perhaps entirely in their floral (or floristic) constitution. The vegetation will be the same in all cases, but the actual kinds of plants which comprise it—and which together compose its flora—will be different. The beech of English woodlands is not the kind of beech which grows in the North American forests, nor do either of these occur in the southern hemisphere, where their place is taken by other related species. The distribution of *floras*, in contrast to that of vegetation, is shown in Fig. 6.

Just as vegetation is chiefly a matter of quantitative characters, so flora is chiefly a matter of quality, in the sense that it concerns the family relationships of the plant life rather than its visual resemblances. The flora of a region is the total of the species within its boundaries, but the vegetation is the general effect produced by the growth of some or all of these in combination.[76]

Plant Geographical Zones

In addition to the units shown in Fig. 6, it is necessary to recognize also the zones of A. Eig[66]—the well-known Saharo-Sindian, bordered on the north by the Mediterranean and Irano-Turanian and on the south by the less known and perhaps less generally accepted Sudano-Deccanian. He later published a map of Palestine which he claimed was an improvement on that of 1931, showing the meeting of his first three types in that country. Eig[67] stated that the best criterion for the demarcation of plant geographical territories was still open to discussion, but he himself believed that the most important criteria were: exact floristic study, knowledge of the geographical distribution and ecological requirements of species and their forms (autecology) and the study of associations and their distribution (synecology). He gave a warning against the use of climatic formulae.

Our interest in the western and central parts of the transect is primarily with the Saharo-Sindian zone, which may be considered subtropical rather than tropical. Eig's term has been adopted by M. Zohary,[262] Th. Monod,[150] J. Lebrun[131] and R. Maire and Th. Monod.[137] H. Heine[87] has noted four representatives of the Saharo-Sindian element in the flora of Senegal and Mauritania (see Charles Rossetti in Chapter IV). H. Jacques-Félix[99] uses Eig's term Sudano-Deccanian for a floral region that would include the Saharo-Sindian element, but apparently also most of the West African savannah or Sahelo-Sudanian zone; the grass *Schoenefeldia* is stated to have a Sudano-Deccanian distribution. One may agree with F. N. Hepper's[89] doubts regarding this zone, particularly when comparing the African zones with the Deccan in India, where *Sehima nervosum* (associated with *Dichanthium annulatum*) is thought by some to be

FIGURE 6

Floristic provinces of the world, redrawn from R. Good[76] by P. Dansereau[60]

Key indicating units that come within the transect

Palaeotropic Kingdom

A. *African Subkingdom*

9. North African-Indian Desert Region
   (*a*) Sahara-Arabia (except the south)
   (*b*) Mesopotamia—South Persia—West Pakistan
10. Sudanese Park Steppe Region
   (*a*) Senegambia-Sudan
   (*b*) Upper Nile-land
11. North-east African Highland and Steppe Region
   (*a*) Abyssinia and Eritrea
   (*b*) Galaland and Somaliland
   (*c*) Yemen and South Arabia
   (*d*) Socotra

B. *Indo-Malaysian Subkingdom*

17. Indian Region
   (*a*) Ceylon
   (*b*) Malabar coast and southern India
   (*c*) Deccan
   (*d*) Ganges Plain
   (*e*) Flanks of the Himalaya
18. Continental South-east Asiatic Region
   (*a*) Eastern Assam and Upper Burma
   (*b*) Lower Burma
   (*c*) South China and Hainan
   (*d*) Formosa and the Ryukyu Islands
   (*e*) Siam and Indo-China
19. Malaysian Region
   (*a*) The Malay Peninsula
   (*b*) Java, Sumatra and the Sunda Islands
   (*c*) Borneo
   (*d*) Philippines
   (*e*) Celebes and Moluccas
   (*f*) New Guinea and Aru

Australian Kingdom

32. North and east Australian Region
   (*a*) Northern forests
   (*b*) Queensland forests
34. Central Australian Region
   (*a*) North and east savannahs

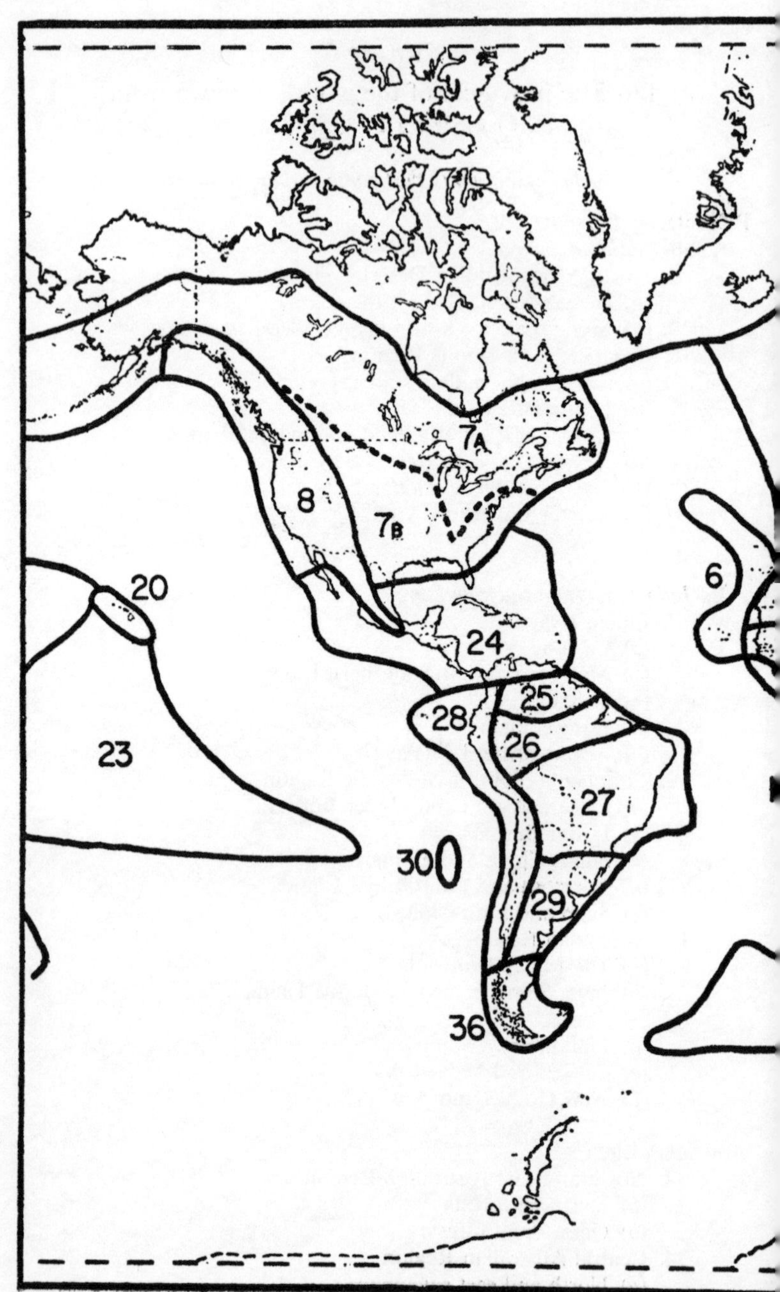

7A
8
7B
6
20
24
25
28
26
23
27
30
29
36

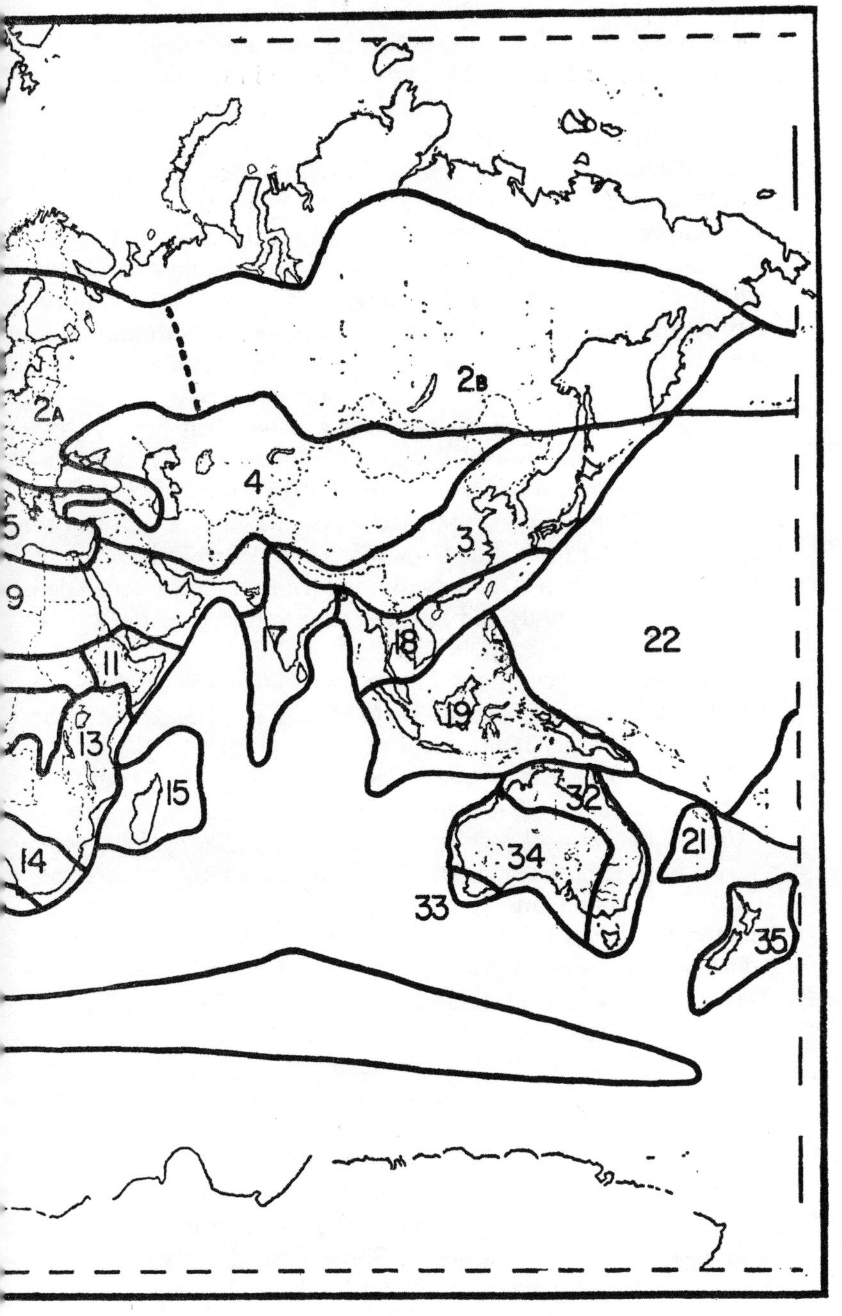

2A
2B
4
3
5
9
11
13
15
14
17
18
19
22
32
34
33
21
35

the top species in the grass succession over the whole of Peninsular India (see *Sehima/Dichanthium* Type in Chapter VI).

We find, however, that M. Zohary[263] later evolved a new interpretation of the situation and of the meaning and scope of terms such as Saharo-Sindian and Sudano-Deccanian. Zohary would change Eig's Saharo-Sindian region to the Saharo-Arabian region, stating that it is clear that the areas of tropical flora in southern Iran, Baluchistan and Sind should be included within the Sudanian region. The climate of Zohary's Saharo-Arabian region is of the Mediterranean type, rainfall limited to the winter season and rarely exceeding 100 mm., summers long, very hot and dry, hence not of a monsoonal type. In north-western Africa, this region occupies a broad belt between the Sudanian and the so-called Mauritanian Steppe (Hauts Plateaux) province. From Cyrenaica eastwards, this belt broadens considerably and almost approaches the Mediterranean Sea to the north (see Figs. 6 and 7 of M. Kassas[109]). Farther east, it occupies the bulk of Arabia (with the exception of Yemen, Aden and the Hadhramaut) and also the southern part of the Syrian desert and lower Mesopotamia.

The region is very poor in species, and has never been an important centre of speciation. Many of its species are clear derivatives of other regions, for example, from:

Mediterranean —*Medicago*, *Lotus*, *Bromus*
Irano-Turanian—*Trigonella*, *Astragalus*, *Stipa*
Sudanian —*Lasiurus*, *Dichanthium*, *Andropogon*, *Panicum*, *Tricholaena*, *Pappophorum*, *Pennisetum*

Zohary's Sudanian region denotes the northernmost plant geographical belt of the Palaeotropic kingdom in Africa and south-western Asia. It coincides only in part with the Sudano-Deccanian of Eig, it being now generally agreed (Good, Gruenberg-Fertig) that the Deccan should be a separate entity. On the other hand, as already stated, some parts of Eig's Saharo-Sindian should be included in Zohary's Sudanian zone, for example southern Iran and Baluchistan. Thus the Sudanian region as understood by Zohary may be regarded as a latitudinal belt extending from the Atlantic coast of Africa along the whole of northern Africa, south-western Arabia, southern Iran and Baluchistan to Sind and Rajasthan. Some French workers do not include within this region the central Sahara, although its vegetation abounds in tropical species. Zohary believes, however, that the northern limit of his Sudanian region should be drawn along the centre of the Sahara, possibly through its most sterile part.

The Sudanian region is distinct from the adjacent Saharo-Arabian region by having hundreds of genera and thousands of species that do not occur in the Holarctis. It is, however, more difficult to distinguish from the next region in Africa south of the Sahara, the Guineo-Congian region. The Gramineae of the Sudanian region include: *Cenchrus catharticus, Cymbopogon schoenanthus, Desmostachya bipinnata, Dichanthium annulatum, Elyonurus royleanus, Eremopogon foveolatus, Lasiurus hirsutus, Panicum turgidum, Pennisetum dichotomum, P. orientale, Tetrapogon villosus, Tricholaena teneriffae.*

M. Kassas[109] refers to a 'perennial grassland form' in his vegetation of the arid lands of north-west Africa, comprising community types dominated by *Lasiurus hirsutus, Panicum turgidum, Pennisetum dichotomum, Poa sinaica, Hyparrhenia hirta* and others, thus bringing together monsoonal and Mediterranean species.

Zohary divides his Sudanian region, which is our monsoonal region as far east as Rajasthan, into western and eastern sub-regions. Although the flora changes with almost every degree of latitude, one may on that basis also distinguish within the western sub-region the Saharo-Sudanian, Sahelo-Sudanian and south Sudanian provinces, and within the east Sudanian sub-region the Nubo-Sindian and the Eritreo-Arabian provinces.

Also relevant to this discussion is the UNESCO project for the standardization of vegetation maps; the vegetation map of Malaysia prepared by C. G. G. J. van Steenis;[232] the second edition of the vegetation map of Africa south of the tropic of Cancer produced on behalf of l'Association pour l'Etude taxonomique de la Flore d'Afrique tropicale (AETFAT) with the assistance of UNESCO;[4] the grass cover of Africa with its associated types of vegetation prepared by J. M. Rattray;[183] and the grass cover of India.[254]

Definition of Grass Cover

Throughout this book, grass communities are frequently spoken of as if they were separate ecological entities and not merely

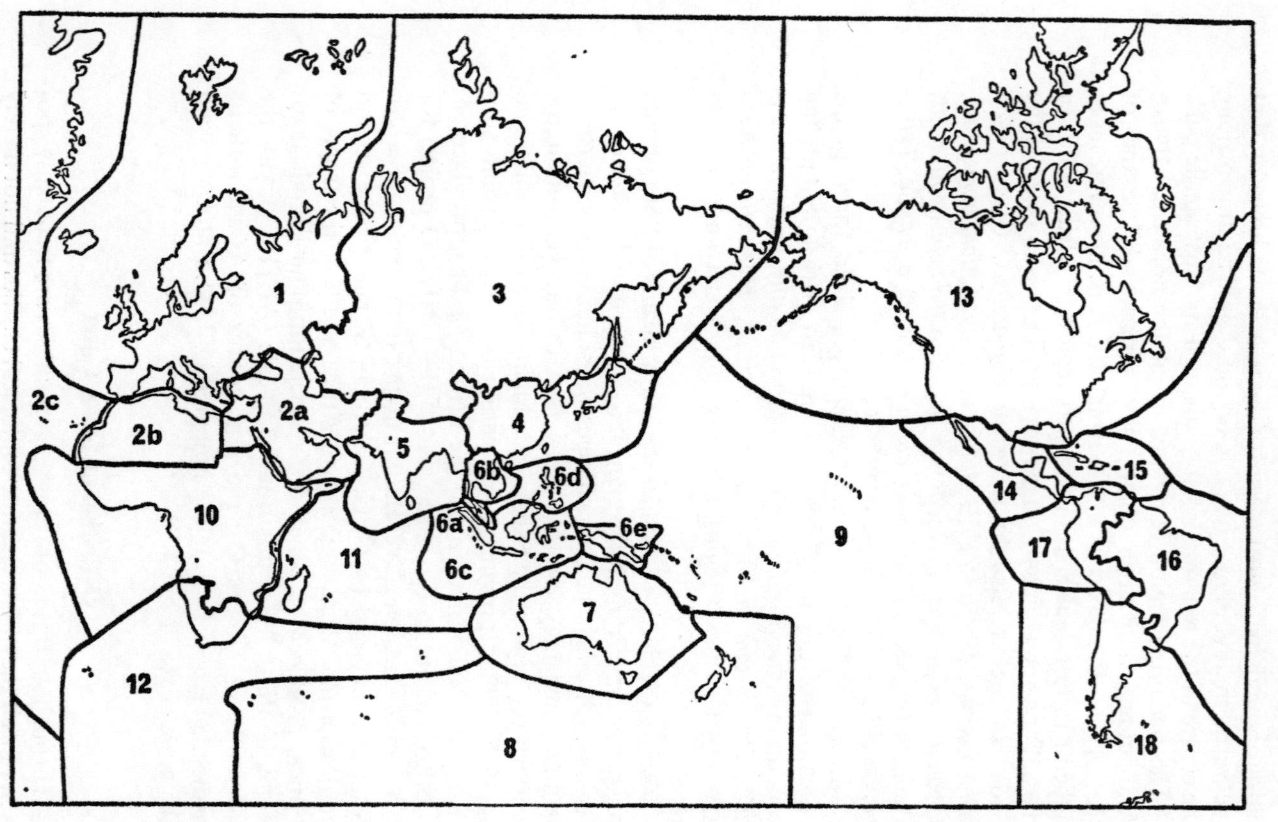
1
2a
2b
2c
3
4
5
6a
6b
6c
6d
6e
7
8
9
10
11
12
13
14
15
16
17
18

FIGURE 7

The subdivision of the world used at the Royal Botanic Gardens, Kew

Key

1. Europe
2. North Africa and Orient (as far as Afghanistan and Baluchistan)
3. Northern Asia
4. China and Japan
5. India
6. Malaysia (including Borneo, Philippines and New Guinea)
7. Australia
8. New Zealand
9. Polynesia
10. Central Africa
11. Madagascar and the Mascarenes
12. South Africa (including South West Africa)
13. North America
14. Central America
15. West Indies
16. East Tropical South America
17. West Tropical South America
18. Temperate South America

part of the total vegetation. Grassland ecologists with a practical bias tend to search for botanical principles and facts that may be of value in the general definition of grass cover types as understood by J. M. Rattray,[183] with particular reference to their place in the natural vegetation, their economic significance to animal husbandry, and the possibility of improvement and better utilization. It is therefore not intended to become involved in complexities of plant taxonomy, nor in the recognition and nomenclature of plant communities and associations on the basis of the ecological principles and tenets applied by different schools.

A type of grass cover is, therefore, a loose term which may be employed and understood by the field officers responsible for managing 'natural' grazing land on the basis of plant succession. All they need is a label for defining the general types of grass cover on the basis of the most 'characteristic' but not necessarily the dominant genera and species. This may then be related to the factors of the total environment in which it occurs, to associated forest type, and so on. It may also be used for purposes of comparison with grass covers in other countries and continents.

Demographic Considerations

Certain paragraphs from *Monsoon Asia* by E. H. G. Dobby[65] provide an excellent background for a consideration of the economic past and future of the most important part of the transect. Of the present world population of 3,355 m., 2,000 m. live in or on the fringes of our monsoonal transect, and by far the greatest proportion of these in Monsoon Asia—see Figs. 8, 9 and 10 from the book by J. Beaujeu-Garnier,[8] and Table 1, showing the true extent of the poverty of the peoples of Monsoon Asia—compared with the gross national product per head in certain advanced countries, taken from ECAFE Reports.

> By Monsoon Asia is to be understood the territory of the countries from West Pakistan through India, Burma, Siam, Indochina and Eastern China as far as the Gulf of Chihli and the off-shore island groups of Japan, the Philippines, Indonesia and Ceylon. The monsoon condition does not, of course, obtain throughout this area, but to use the boundaries of these countries as its limit has statistical convenience. Monsoon Asia is thus the southern and eastern part of Asia, and its

Table 1

The extent of the poverty of the peoples of Monsoon Asia

Gross national product per head			Average annual rates of growth of gross national product, population and income per head (in per cent)			
Country	U.S. $	Population (1965) (millions)	Period	G.N.P.	Population	G.N.P. per head
U.S.A.	2,790	194·6	1951/63	3·1	1·7	1·4
New Zealand	1,617	2·6	1954/63	3·9	2·2	1·7
Australia	1,533	11·4	1950/51–1964/65	4·1	2·2	1·8
France	1,406	48·9	1951/63	4·6	1·1	3·4
United Kingdom	1,361	54·8	1951/63	2·6	0·5	2·1
Japan	589	98·0	1951/64	9·2	1·1	8·1
Total population		410·3				
Burma	65	24·7	1951/64	4·8	2·0	2·6
Cambodia	111	6·3	1952/63	5·8	3·6	2·2
Ceylon	129	11·3	1951/64	3·1	2·5	0·5
China (Mainland)	56	764·0	—	—	—	—
Hong Kong	305	3·8	—	—	—	—
India	78	483·0	1951/64	3·9	2·1	1·8
Indonesia	127	105·0	1951/60	3·3	2·1	1·1
Korea, Republic	113	28·0	1953/64	4·9	2·4	2·3
Malaysia	222	8·4	1955/64	4·8	3·2	1·6
Nepal	67	10·0	—	—	—	—
Pakistan	86	113·0	1951/64	3·5	2·1	1·3
Philippines	134	32·0	1951/64	5·4	3·1	2·2
Singapore	449	1·9	—	—	—	—
Taiwan	154	12·4	1951/64	8·1	3·5	4·5
Thailand	104	30·6	1951/64	6·1	3·0	2·9
Viet Nam Republic	97	16·0	—	—	—	—
Total population		1,650·4				

Figures for gross domestic product per caput relate to 1963, and are taken from the *Economic Survey of Asia and the Far East* (1965) published by ECAFE, Bangkok, with the exception of Mainland China and Indonesia, which figures are taken from Jagdish Bhagwati, *The Economics of Underdeveloped Countries*, Weidenfeld and Nicolson, London, 1966, and relate to the mid-1950's. Figures for annual growth rates are taken from the above ECAFE *Economic Survey of Asia and the Far East* (1965).

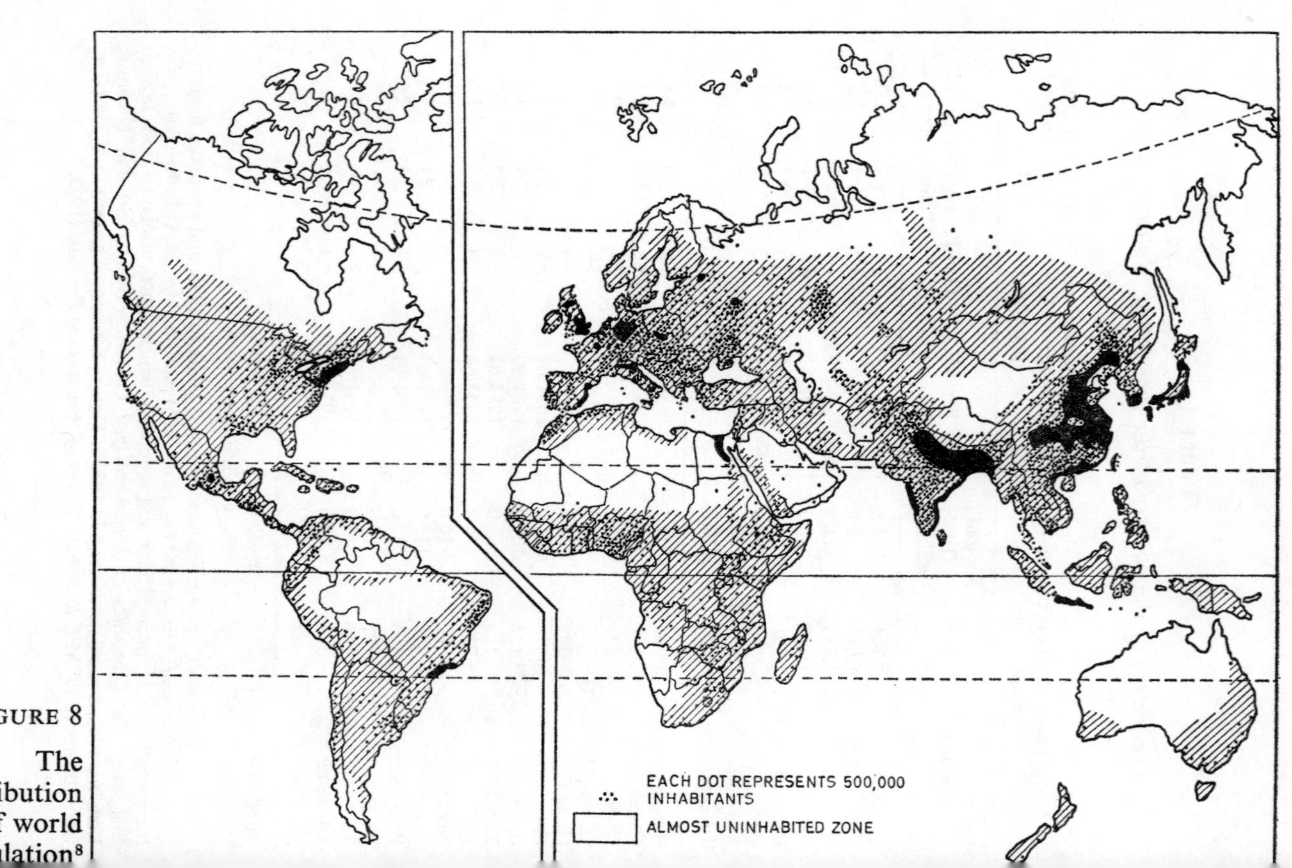

FIGURE 8

The distribution of world population[8]

features have been described as 'the golden fringe to a beggar's mantle' and as 'the golden crescent'. It includes less than a seventh of the world's land surface, yet contains half of humanity and nearly a third of the world's cultivated land, ranking as the most important of regions.

The world's oldest civilisations (of North India and North China) are identified with Monsoon Asia, their very antiquity being an element in the complexities of problems arising there today. In speech, colour and physique the people differ considerably from those of surrounding regions; they also differ greatly from place to place inside their 'Oriental World'. Among millions of them, dress, food habits, domestic manners, social customs and modes of farming remain much as they have been for millennia, and largely peculiar to the Orient. The self-contained peasant and the rural way of life in small villages persists and forms one of the major contrasts with the Western World, where urbanisation, commercial farming and industry set the characteristic pattern. Its landscape is dominated by cereal farming in contrast to W. Europe, where grass covers much of the used land and is the basis of animal farming. Cotton is the usual textile fibre in the East, wool in the West. Mechanisation sets the standard elsewhere, manual methods are still a feature of the East. The West is a world of metals in wide and everyday application: Monsoon Asia is the home of a 'vegetable civilisation', where wood and plant materials have been the basis of everyday things from houses to domestic utensils, farm implements and clothing.

Problems chiefly affecting Asians domestically are: How can the age-old practices of the rural areas be brought into line with the modernity of Asian cities and industrial centres? Whence can come the food supplies for the urban areas if rising standards of diet and new forms of land tenure in the countryside make surpluses things of the past? Can the pressure of population be absorbed by domestic developments alone? Is it possible for the plural societies to be welded into nations? Are the gulfs between the ancient and the modern aspects of Asian economic life bridgeable?

Because the geographer is concerned with material things and their associations on the landscape, both locally and re-

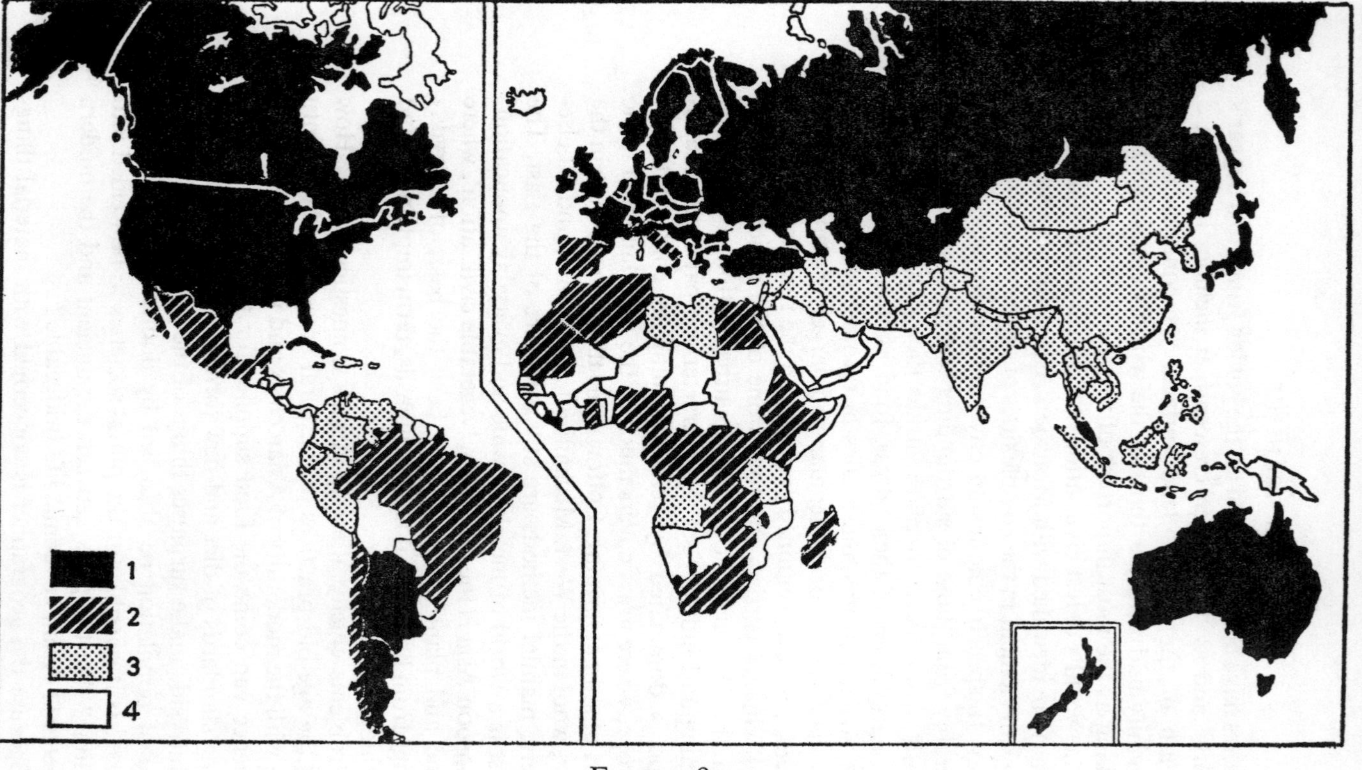

FIGURE 9

Number of calories per person per day, according to FAO estimates made in 1962

1. More than 2,700 2. From 2,200 to 2,700 3. Less than 2,200 4. No information available

gionally, he has a special contribution to make in appraising these issues. His kind of survey of actualities in Asia is a vital preliminary to development and planning as well as to understanding. His conception of the association and relationship between human needs and natural environment is a corrective to those who would introduce a change without carefully estimating the displacements and replacements it will involve in the region's human ecology.[65]

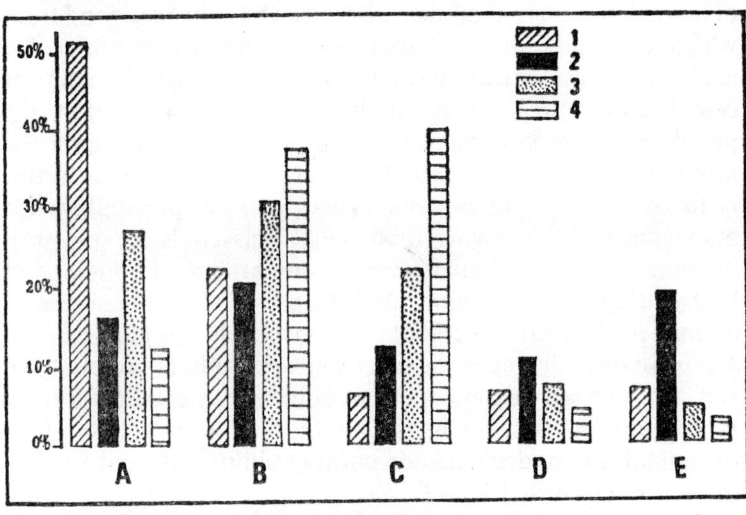

FIGURE 10

Comparative graphs (1962) of population, cultivated area, production and revenue in the different regions, according to J. Beaujeu-Garnier[8]

A. Far East
B. Europe
C. North America
D. Latin America
E. Africa

Scale in percentage of world total

Economic Evaluation of Monsoonal Grasslands

Although much of this book is concerned with data which may appear to be rather academic, in this day and age it is essential to look for the practical and economic applications.

For the first time, the location and nature of the grasslands or grass covers or grazing lands of one of the major ecoclimatic zones of the world are reviewed in as detailed a manner as space and available data will permit. Because the grasslands of this zone are poor in yield and/or protein content, high in fibre and seasonal in availability and palatability, they provide a very inadequate basis for the nutrition of productive animals. Thus the human populations of this zone suffer from serious or chronic malnutrition due to a general lack or complete absence of animal protein in their diets. Because the cultivated land from which feeds and fodders for milk (and poultry) production come is already more than fully occupied in meeting the demand for cereals and grain legumes for direct human consumption, the people in the monsoonal lands will have to depend upon the humid temperate environments for their animal protein, if this is to be in the form of milk. There may be possibilities for managing and improving monsoonal grasslands for the production of sheep and goats (under semi-arid conditions), or of beef cattle (under medium rainfall), provided these sources of animal protein are acceptable to the people. In Chinese and Japanese diets the pig is the chief source of animal protein from the land—in fact in some Chinese languages the word for meat is the word for pork. But here again the maintenance of this type of animal husbandry depends entirely upon feeds and fodders grown on cultivated land.[271]

There is nevertheless a tendency among workers in the tropics to consider that tropical grasslands represent a vast untapped resource for the production of livestock, as witness the recommendation proposed by G. S. Puri (Ghana) and seconded by J. A. Richards (Jamaica) at the 9th International Grassland Congress in Brazil in January 1965:

Whereas:

(1) The tropics have vast unexploited potentialities for higher animal production through improvement and better management of the grasslands; and

(2) That rapid development towards this potential demands the best possible orientation and coordination of research and other activities in this field; and

(3) Considering FAO's wide responsibilities and programmes for better grassland production and use:

> This Congress recommends that FAO establish an International Tropical Grassland Commission to coordinate research in tropical pastures, suggest appropriate priorities and promote promising activities and cooperative efforts among the tropical countries in pasture research and development.

A meeting of the 10th International Grassland Congress was informed by R. A. Peterson (FAO) that the FAO Conference had recommended to the Director-General that a Tropical Grassland Advisory Group be established, to focus attention on the importance of tropical grasslands and to promote their development.

Another recommendation at the 9th Congress referred to the practicability of establishing world zones or regions, specifically in this case for planning the rotation of future Congresses, but which would also be of value in planning Grassland Congresses on an ecological rather than a subject-matter basis. At the 10th Congress, the Executive (Continuing) Committee made no recommendation on the establishment of world zones for the rotation of future Congresses, but merely requested the Committee to 'take cognizance of the sites of previous Congresses and of the major geographic areas of the world'. The Committee is to contain one representative from each of the following geographic areas:

- (i) U.S.A. and Canada
- (ii) Latin America and the Caribbean
- (iii) Australia and New Zealand
- (iv) South-east Asia
- (v) East Asia
- (vi) Mediterranean area and Near East
- (vii) Europe (excluding vi)
- (viii) Africa (excluding vi)

It is hoped, however, that this book will indicate the limitations of the geographical or regional political approach, and that it will indicate the possibility, and more the urgent need to recognize and define the macro-zones of the world's grasslands in relation to ecoclimatic conditions. Current work on the planning of animal husbandry and development for the maximum production of animal protein in one form or another must necessarily be based upon the correct assessment and development on a correct ecological basis of the grassland and fodder resources of countries and geographical regions. It is the climate, the growth form of the grasses and above all their seasonal pattern of growth that define the type of animal husbandry that may be maintained or established. What an exciting Conference it would be, and what great scientific and practical progress could be made, if we could bring together all the workers on monsoonal grasslands.

When an economic evaluation of present use and future potentialities of the monsoonal grasslands is made from the point of view of the animal agronomist and not of the plant ecologist, crop agronomist or forester, it will probably be found that tropical grasslands in general, or monsoonal grasslands in particular, are somewhat of a snare and a delusion. They have no potentiality for the production of milk, and probably only for the less intensive and economically efficient forms of production of beef and mutton. It is time that the aura of sanctity were removed from the words 'grasslands' and 'pasture' and that they should become the subject of a realistic reappraisal of their value in an overall plan for optimal use of land.

Any discussion of the economic value of monsoonal grasslands should consider methods of improving them. Ecological management, by which is meant protection to promote regeneration of cover, followed by controlled grazing, is the method most suitable for the lower rainfall parts of the zone. One may also have assisted ecological management involving the introduction into a degraded stand of the superior species in the same succession by surface or strip seeding or the planting of mother plants on the edge of contour furrows. Methods of water conservation must also be studied. Again in the arid and semi-arid parts of the transect, it is necessary to study and map routes of transhumance in systems of nomadic and migratory husbandry in order to be able to plan improvements intelligently. Finally there are all the complex relations between climatic factors and animal physiology and production in tropical and sub-tropical areas.[136] One should stress the need for intensification of systems of animal husbandry wherever possible. There is little enough natural or even superior disturbed vegetation left in the world, and its further destruction by the ever greater spread of extensive systems of grazing of low-productivity animals cannot be supported by the conservationist or the forester.

In an economic assessment of the potentialities of monsoonal grasslands it must be noted, however, that fabulous yields of nutritious green feed can be obtained from certain species of grasses grown on cultivated land. In fact, if these cultivated grasses are fed through reasonably productive milch bovines, the cash return per unit of land is likely to be higher than that obtainable from any other food or cash crop. In India it is usually stated that yields in an all-cutting zero-grazing regime of the order of 100,000 kg. green per ha. may be obtained from the African grasses Guinea, Napier, Napier × finger millet hybrid, and Para, grown with clean water irrigation plus nitrogenous fertilizer (40 kg. N per ha. after every second cutting); 200,000 kg. green per ha. when irrigated with cowshed wash, and 300,000 kg. per ha. when irrigated with city sewage which has been correctly treated by sedimentation and diluted, and then distributed on waste land. Obviously it is not correct to consider grazing such highly productive grass stands since this would involve the loss of 40 to 50 per cent of the yield through livestock trampling and poisoning of the herbage by dung and urine. Where demographic pressures are not so great as to make it impossible, some more conservative workers with a European background are attempting to devise systems of grazing of these types of grasses or of others which have a more suitable growth habit for grazing and therefore a lower yield. The economic viability of such a system under monsoonal growing conditions with any superior type of bovine not too resistant to exposure in the heat of the sun still needs to be proved, and compared with the cut-and-carry system combined with a high degree of mechanization of a specialized type. In the zones of lower rainfall some degree of success has been achieved with the reseeding of adapted grasses for use by sheep, goats or cattle of low productivity or which are being bred for draught purposes, but at least in the Asian part of the transect, if land can be ploughed up, first consideration must always be given to the prior claims of food production for direct human consumption.

While considering domestic animals, reference should also be made to the current interest in wildlife on marginal agricultural land in Africa. H. P. Ledger[132] has shown that in East and Central Africa game can produce more and cheaper meat per unit area with less destructive effect on vegetation and soil than domestic livestock, because of higher efficiency of feed utiliza-

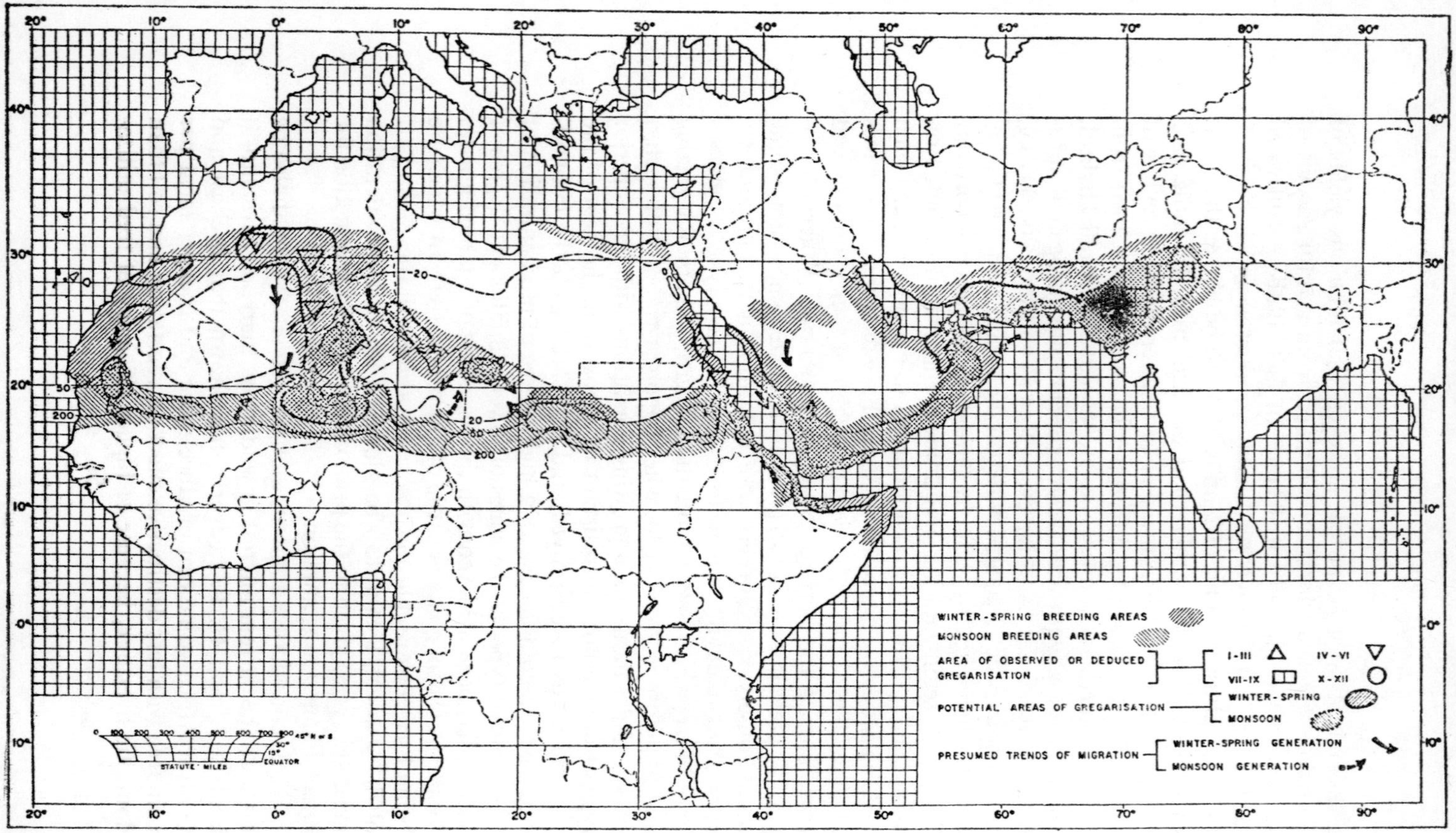
WINTER-SPRING BREEDING AREAS
MONSOON BREEDING AREAS
AREA OF OBSERVED OR DEDUCED GREGARISATION
I-III
IV-VI
VII-IX
X-XII
POTENTIAL AREAS OF GREGARISATION
WINTER-SPRING
MONSOON
PRESUMED TRENDS OF MIGRATION
WINTER-SPRING GENERATION
MONSOON GENERATION
STATUTE MILES
EQUATOR
20
50
200

FIGURE 11

tion, lower water requirements, greater mobility, greater rates of growth and reproduction and higher resistance to disease. Well-planned game-ranching schemes may provide a useful stepping stone from a subsistence to a productive economy, by creating buffer zones containing only game between the true desert and the good ranching lands. Grazing pressure by wild animals results in pasture rejuvenation which is manifested by the development of a grazing mosaic.[238] A sequence of animals, heavier ones followed by lighter ones, use the different pastures in rotation during the year. Thus there are alternate periods of optimal use and rest, and the harmful effects of overgrazing are avoided. Dry as distinct from extremely wet periods appear to favour the fauna.

Impact of Man

The activities of man have had a great impact upon the total land resources throughout the monsoonal transect. This has been the subject of several international meetings, for example, the 9th Technical Meeting of the International Union for the Conservation of Nature, dealing with the impact of man on the tropical environment in Africa (Nairobi, September 1963), and the Symposium on the impact of man on humid tropical vegetation held in Papua and New Guinea in September 1960.[1] In matters of human ecology, it has to be admitted that the transect in general and its arid and semi-arid western parts in particular represent a most inhospitable environment in which to make one's living from the land. And in Africa in particular, the biome is further complicated by the coincidence of the distribution of the desert locust (Fig. 11), and of the *Quelea* or weaver birds, which subsist on the seeds of the wild grasses and cereals of the zone (Peter Ward in *Geographical Magazine* for July 1966).

This book has been confined to a review of the information available on the grasslands in monsoonal environments, in which two-thirds of the world's population lives. It should be followed by similar studies within the same ecological zone, for example, on the classification of land for the planning or re-planning of optimal land use, or pre-investment surveys; on the types of farming and cropping systems that are at present practised or may be introduced, without or with irrigation; on the types of animal husbandry that exist now or may be developed.

Chapter II

THE MONSOONAL ENVIRONMENT

We have seen that, on a population basis alone, the lands lying within or bordering upon the monsoonal transect should be given priority in the joint studies of the environment made by ecologists (plant, animal and human), meteorologists and climatologists. The monsoonal environment, with its dual hazards of excess and deficiency of rainfall, is the most difficult environment in which to farm well and with ensured economic success year in year out. Full marks must be given to all those farmers who live and make a profit or at least a reasonable subsistence for their families under conditions where practically all they have to depend on is what L. P. Smith has called their 'peasant wisdom' (*Nature* of 10th September 1966).

A very small proportion of existing knowledge is correctly applied to practical agriculture on a world scale. Errors are still being made in the use of land that were made by the Romans 2,000 years ago. But we can no longer rely on the hit-or-miss, trial and error methods. Haste is inevitable to obtain quick answers to land use problems, based on more research on the micro-scale, while making more use of available macro-data. Agroclimatology has so far been academic rather than practical. Simplicity has often demanded and been granted the sacrifice of validity. 'While the world and world science owes much to its enthusiasts and non-conformists, the energetic worker with the second class mind can cause infinitely more trouble than the gifted idler' (L. P. Smith).

Not being at the time fully aware of the prior and critical claims of monsoon Asia, the first FAO/UNESCO/WMO joint project on agroclimatology which I was able to establish some years ago was concerned with the semi-arid, winter-rainfall countries of Lebanon, Syria, Jordan, Iraq and Iran. An extensive Technical Report was prepared by the specialists concerned,

G. Perrin de Brichambaut and C. C. Wallén,[168, 240] in mimeographed form for distribution to the institutions and departments concerned in member countries. WMO later issued an abbreviated version as Technical Note No. 56.

The second survey by the same Agencies, still in progress in 1966, by J. Cochemé and P. Franquin, has been concerned with the far western part of the transect, defined by Cochemé[48] as 'a tropical area with a semi-arid climate of summer rains in Africa south of the Sahara, on the belt bounded to the north by arid steppe and desert and to the south by humid savannah and forest . . . the northern limit would be that of dryland farming corresponding more or less to the isohyets of 400 to 500 mm. of mean annual rainfall, the approximate southern limits would be where the length of the dry season exceeds that of the wet season' (see also below, page 74 and Appendix 1).

The results of the first survey had little application to the conditions of monsoon Asia—the techniques adopted may be of interest. The second survey is admittedly in an area of true or pseudo-monsoon, following P. Pédelaborde,[166] but again has little application to the real problem areas of dense human populations and critical nutritional deficiencies of monsoon Asia. It is sincerely to be hoped that the three U.N. Agencies will devote a third survey to this region. A team of specialists should work at some appropriate centre of research in the region, not in Europe. In addition to the factors considered in the first two surveys, they should make special studies on expectancy of crises. Politicians and administrators in the region fall back too readily on 'abnormal weather conditions' to explain their failures in planning, which are so frequently exercises in wishful thinking, based upon figures for the good years. It must be realized that there is no such thing as abnormal weather—it is the expectancy of critical years of surplus or deficit rainfall in what is a normal weather cycle, characterized by a certain degree of irregularity ranging between extremes. If we know that a critical year or season is likely to occur once in ten or once in twenty years, plans for the accumulation and careful storage of grain can be made accordingly.

L. P. Smith has written (*Nature* of 10th September 1966) that assessment of climate is a factor in determining correctly the economics involved in seasonal decision-making in practical agriculture. It is often thought that while weather is variable, climate is constant, but nothing could be more misleading. Unfortunately climatic changes or trends are rarely identified in meteorology until they have almost ceased to be effective in terms of agriculture, or agriculture may have become adjusted to the new conditions before scientific guidance can be given. We do not know whether a monsoonal climate is more or less susceptible to change than any other type of climate.

In addition to agroclimatology and meteorology, the other factors of the physical environment to be studied include topography, geomorphology, geology and soils, in so far as they together govern vegetation communities and the adaptability of plants. Acting upon these are the activities of man, from the destructive ones that cause devegetation and abnormal erosion, to the constructive ones such as the flooding of forested valleys for the purpose of reservoirs.

Over much of our transect, particularly in its western part from the Atlantic Coast to India, the actions of man in his use and misuse of the total land resources play a dominant role, affecting to a greater or lesser extent the nature and the degree of the expression of the factors of the physical environment. We find in the chapters on Taiwan and Papua and New Guinea the great effects of burning and/or shifting cultivation on the grass covers, and on their value as indicators of potential land use. In these regions, the natural vegetation has long been exposed to a steady increase in intensity of unplanned, uncontrolled and therefore destructive use of the land. Much of what is described below as the grass cover component of the vegetation is for these reasons low in the ecological succession for each particular habitat, except in areas remote from human influences.

The site data given in Part 2 of *The Grass Cover of India*,[254] for example, show the effects of excessive use on former grass-dominant associations that were higher in the ecological succession. Anywhere within daily walking distance of a village or where free-ranging nomadic and migratory flocks and herds move in large numbers, it is the lower stages in succession that are to be found, and it is sometimes difficult to recreate the hypothetical climatic climax (or rather sub-climax in a forest or woodland climax). In planning for ecological management, we generally know what should be done to raise a degraded community to a higher level and to maintain it there. As in so many parts of the tropics and subtropics within and outside the transect under review, we cannot put that knowledge into effect because of human, social and political factors beyond our control.

Definition of Monsoonal Ecoclimate

An ecoclimate may be defined as the climate of any environment in which plants and animals live and have their being. The fact that grasses such as *Panicum turgidum*, *Lasiurus sindicus* (or *hirsutus*) and others are to be found throughout that part of our transect that extends from the Atlantic Coast in West Africa to Rajasthan in India would indicate to the plant ecologist that there must be some factor or factors which operate to create a similar environment. If that environment is monsoonal at the eastern end in West Pakistan and India, why should it not be monsoonal also at the western end? The fact that meteorologists and agroclimatologists are still struggling to define a monsoonal environment seems to show that they do not take the indicator value of plant communities and their component genera and species adequately into account.

The word *monsoon* has come to be used in a number of ways, referring to a type of wind, to a rainy season, and to violent rain during that season. E. H. G. Dobby[65] defines the history and modern usage of the term as follows:

> Arabs first used their expression 'mausin' to mean a season of winds such as occur around the Arabian Sea where for about six months a year they blow as northeasterlies and for the other six months as southwesterlies. Arabs thought of seasons in terms of wind because there were no other significant changes through the year in their warm, arid climate—as natural as for Europeans to think of seasons in terms of varying warmth and cold, variations which set the regime of growth in their part of the world. It is equally natural for people within the tropics to think of seasons in terms of raininess, variations of which dominate their farming. The prolonged association of seaborne Arabs with South and East Asia explains why their word and conception have been accepted and perpetuated all the way from the Red Sea to Japan. The Chairman of the WMO Working Group on methods of

analysis and forecasting in the tropics has proposed the following definitions:

Thermal Equatorial Trough (TET): This is a zonal trough at the surface of the earth or immediately above it, lying along the belt of mean daily maximum surface temperature (thermal equator). It moves annually according to the inclination of the sun and is therefore chiefly confined to tropical regions. In an approximately zonal distribution of land and sea (e.g. the coastline of South Arabia and South Asia) there may be a dual phenomenon, as a thermal trough may form over the mainland in the summer hemisphere, while the trough persists over the sea.

Thermal Equatorial Convergence Zone (TECZ): This is a convergence zone for winds produced by the TET, the converging winds usually coming from both hemispheres. This zone lies in the TET when the thermal equator is near the geographical equator, but moves towards the latter when the thermal equator is far away from the geographical equator (solstice position over land). In the first case the converging winds are the NE and SE trades, while in the second case, monsoon winds from the winter hemisphere and westerlies converge towards the equator side of the TET.

Inter-Tropical Weather Zone: This is a zonal concentration of cloud and meteorological systems, mainly convective in the summer hemisphere within the TECZ or displaced nearer the equator. The ITWZ may move equatorwards away from the TECZ for several hundred kilometres above the mainland during the solstice period. This is due to the fact that at this time the TECZ is often covered by subsident air masses preventing cloud development despite maximum surface convergence. On the other hand, kinematic conditions are extremely favourable to cloud formation nearer the equator where this is not hampered by upper-air subsidence, despite weaker surface convergence.

An excellent review on *Les Moussons* has been published by P. Pédelaborde of the University of Toulouse,[166] who has indicated an approximate location of the zones of intertropical pressure in July and January in a map adapted from an earlier one by H. Riehl (Fig. 12). While working at the Institut français in Pondicherry, India, J. Legris[134] summarized the various explanations that have been proposed for the working of the monsoon, introducing also the worldwide dynamic theory.

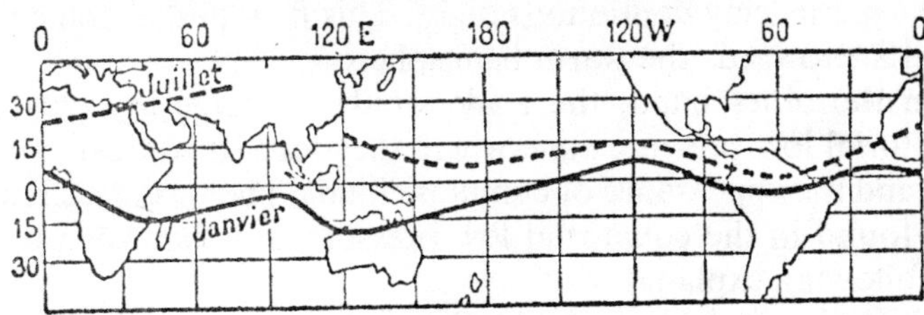

FIGURE 12

Mean position in January and July of the axis of the equatorial low pressure system or Intertropical Convergence Zone, from P. Pédelaborde[166] after Riehl

(1) The thermal theory explains the arrival of cool air from the ocean by the low pressure over the continent during summer. In winter, on the other hand, the cool continental air moves south and south-east.

(2) The theory of the tropical meteorologists gives the name monsoon to 'any air flow crossing the equator and subject to deflection, as opposed to the trade winds which do not cross the equator and are not deflected' (I. Eaker, quoted by Pédelaborde).[166]

(3) According to Flohn, the monsoon is the result of the normal interplay of the entire air circulation, the southern trade wind being confined to the southern hemisphere during the summer. There is simply an expansion of the equatorial low pressure zone, corresponding to equatorial west winds.

These theories do not take account of all the monsoonal phenomena. Recent work shows the relation which exists between general air dynamics and the evolution of the equatorial low pressure zone which creates the monsoon. The worldwide dynamic theory, according to Legris, provides a satisfactory explanation for the Indian monsoon. The extreme limit of the polar zone is characterized by a stratospheric jet-stream, the position and speed of which are determined by the speed of polar circulation. The polar western air flow is deflected towards the equator and its movement is accelerated over the hemisphere during winter. This is explained by the strength and frequency of cold air streams coming from the polar region, which is at that time experiencing continuous night. This flow pushes the equatorial air towards the warm hemisphere.

On the other hand, the trade winds converge towards the equatorial low pressure zone where they lead to vertical expansion and the appearance of cumulus clouds. The western circulation found in the equatorial low pressure zone has been given the following explanations:

(1) Trade winds deflected by the change of hemisphere.
(2) Deflection of the trade winds by the suction of the thermal low pressures over the hot continents.
(3) Cyclonic curves creating a western current on their equatorial front.
(4) Convection movements due to the descent of the trade wind air dilated at high altitudes.

This western circulation can be superficial, as in the Central Pacific, or may reach a height of 4–5,000 m. over the Indian Ocean. This inter-tropical convergence zone (ITC) is characterized by marked cloudiness. It is an area of instability along which cyclonic depressions form and play an important part in the explanation of certain Indian climates.

These general factors, which apply to all monsoon climates, are strengthened and amplified by the topography of the Indian region, and particularly by the obstacle which the Himalayan mountain barrier creates for the seasonal movements of the polar front and the ITC. This is the fundamental characteristic of the Indian monsoon: the mountains hamper the latitudinal zonation of climates; the importance of their influence will appear in the detailed study of climatic factors.

The Monsoonal Ecoclimate in India

The seasonal worldwide movements of the polar front and the ITC are particularly marked over the Asian continent.[130] The polar front varies from 50° N in summer to about 30° N in winter. The mean position of the ITC is about 10° S in winter and more than 40° N over China in summer. Within this general Asian system, the obstacle constituted by the Himalaya and Tibet influences upper-air movement and enhances the effects of the worldwide fluctuations.

During the winter, India is protected from the effects of the

Siberian anti cyclone by the Himalayan barrier. It is therefore a general westerly current which passes through the Gangetic Plain. It arises from the anticyclone over Russian Turkestan and is accompanied by some disturbances leading to slight precipitations on the outer ranges of the Himalaya. These disturbances weaken rapidly and hardly ever reach beyond the longitude of Kathmandu.

This is the region of the subtropical (bixeric) climates (with two dry seasons) found in the Punjab, Northern Uttar Pradesh and the western part of the Nepali Terai. These cold currents lead to a distinct fall in temperature. If they continue for sufficiently long they may have a disastrous effect, as was the case in December 1961, when more than 200 people died and much livestock was killed in Delhi and the Punjab. This current curves eastward and joins the north-east trade wind which blows over the rest of the peninsula and carries very little moisture (winter monsoon). During this period, the high-altitude jet-stream passes to the south of the Himalaya.

During the summer, the ITC covers the whole of India. In April/May, a low-pressure zone at ground level develops over north-west India and West Pakistan. A sea wind arises, which is not strong enough to start the monsoon, but which brings humid air over the continent. This arrival of air masses from the sea explains the high relative humidity recorded even in semi-desert or mountain areas.

In June, due to a slowing down of polar circulation, the polar front moves northward and the jet-stream suddenly switches to the northern side of the Himalaya and Tibet. The low-pressure zone (ITC) suddenly moves over India and brings torrential rains. The south-westerly winds are strengthened by an extension of the southern hemisphere trade winds which may at that time cross the equator and curve eastward. The winds entering the Bay of Bengal are deflected to the north, then the north-east, by the mountains of Burma and the Himalaya. This explains why in the lower Gangetic Plain the rains often come from the south-east.[265]

Comparison with Africa

If one compares this with West Africa and Madagascar, it will be found that the mechanisms and the scope of the phenomena are different. These are not true monsoon climates (according to Pédelaborde). In West Africa, the annual rainfall cycle is similar to the Indian one. But Pédelaborde[166] quotes two great differences from the Asian climate: the worldwide north/south movement is not very great. 'Drought continues over the Sahara north of the 20th Parallel, while the Indian and Malay monsoons reach 30° or 35° N.' Moreover, the cold air from the temperate zone does not reach the inter-tropical belt regularly in winter. Pédelaborde calls this climate a pseudo-monsoon.

The rainfall patterns of Madagascar and Mozambique recall those of monsoonal Asia. 'Rains are plentiful in summer (November to April) and very slight in winter (May to October). The greater rainfall in summer is due to convergence (ITC) and not to a higher degree of humidity of an air mass which is warmer than that in winter. Trade-winds and low pressures alternate here with hardly any change in temperature, as is the case in the purely tropical climates.' Pédelaborde concludes: it may be claimed that here there is a monsoon trend, but only during summer.

Bioclimates of Africa and India

John Phillips[171, 173] has given his own bioclimatic regional classification for Africa, in which he defines the terms: tropics, hotter subtropics, bioclimatic region, forest, wooded savannah, scrub and grassland. Following a discussion at a C.C.T.A. meeting in Yangambi in 1956, A. Aubreville also attempted to produce definitions that would be internationally acceptable to workers throughout Africa. Also relevant are the AETFAT map of the vegetation of Africa south of the Sahara,[4] the climatic map produced for C.C.T.A. by the University of Witwatersrand[100] and J. M. Rattray's map of the grass cover of Africa in relation to the general physiognomy of the vegetation.[183]

The followers of H. Gaussen working in Toulouse and Pondicherry (India) have made detailed studies of the bioclimates of that section of the transect covered by India and south-east Asia. The three main factors selected for the application of this method are:

(1) Mean temperature of coldest month:
 above 20° C

between 15° and 20° C
between 10° and 15° C.

These three divisions coincide with an altitudinal zonation in south India and Ceylon.[135] The isothermal line 20° C for the coldest month follows roughly the 900-metre contour, while the isothermal line 15° C follows the 1,500-metre contour. It is around the latter level that it becomes possible to replace the indigenous monsoonal grass communities with introduced 'European' species.

(2) Total annual rainfall in four classes:
more than 2,000 mm.
between 2,000 and 1,500 mm.
between 1,500 and 1,000 mm.
between 1,000 and 500 mm.

(3) Duration of the dry season, in six classes, according to number of dry months:
none
one to two
three to four
five to six
seven to eight
more than eight.

The approach of this school has been summarized by V. M. Meher-Homji[142] in his review of the bioclimates of India in relation to the vegetational criteria (Fig. 13). 'The purpose of a climatic classification is to characterize the climatic regions in terms of meteorological elements like temperature and precipitation, which are the most decisive in determining the landscape of the earth. The plant-climate relation is so close that it is possible to differentiate the types of climates such as equatorial, tropical, mediterranean, temperate, polar, with their different shades such as perhumid, humid, sub-humid, semi-arid, arid, hyperarid, using the characters of the plant cover. For example, in a physical classification of climate, an arid or a humid climate is defined on the basis of the annual precipitation, or the number of dry months and dry days in the year or by means of several climatic formulae.' Bioclimate is defined as the climate in relation to life and is expressed by an index of aridity-humidity, which combines the three essential ecological factors, temperature, precipitation and dry period (Table 2). Correlation may be made be-

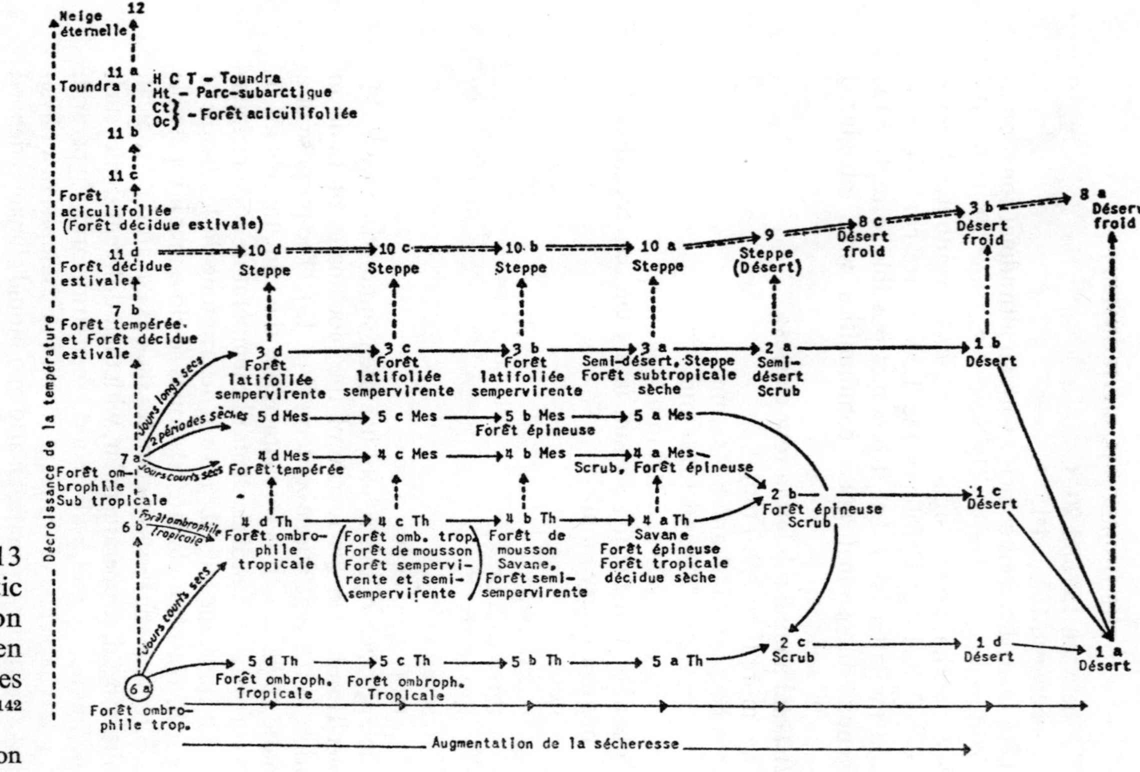

FIGURE 13
Diagrammatic presentation of relation between bioclimates and vegetation[142]

La relation diagrammétique entre les bioclimats et la végétation

tween this index and four distinct criteria adopted by botanists, sociologists and agronomists:

Floristic: The fact that there are certain species or genera or families that are peculiar and restricted to arid or humid regions, and that may be used to characterize these regions, for example Saharo-Sindian, Sudano-Deccanian, Malayan, etc.

Vegetational: Similar climates induce the emergence of similar physiognomic types of vegetation in widely separated parts of the world, e.g. rain forest, deciduous forest, thorn forest, desert, steppe, savannah, tundra, etc.

Morpho-ecological: This is manifested by adaptation to climates and by the prevalence of certain growth forms or life forms under the influence of climate.

Agronomic: A region may be defined in relation to agronomic potentialities.

Ecoclimates of Monsoon Asia

There are several other descriptions of the climates of Asia which relate more to east and south-east Asia (see Figs. 4, 5). At this point it should be stated that we may find monsoonal conditions over most of the region for the whole year, the characteristic pattern of winds and rainfall with minimum temperatures not falling below 0° C. In the north conditions are different. In northern India, north of the Nanling in China and over much of Japan the summer may be monsoonal, but the winter not. Under such conditions one may still grow the dominant monsoon crop, paddy, in the summer, even as far north as Sapporo in Hokkaido. With the introduction of cold winters into the climatic pattern, the typical monsoonal perennial grasses cannot grow, and one finds others adapted to the annual climatic pattern of hot monsoonal summers and cold temperate or even subarctic winters.

According to L. D. Stamp,[212] the three essential elements in the climate of Asia are the dry outblowing winds of winter, the wet inblowing winds of summer, and the controlling influence of the relief. He recognizes three major tropical climatic types, among ten in all:

Equatorial, extending roughly 5° on each side of the Equator.

Tropical Monsoon, mainly within the tropics.

TABLE 2

Annual rainfall correlated to physiognomic features, according to G. Budowski,[22] with particular reference to the Americas

Total Annual Rainfall (mm.)	Natural Vegetation	Height	Crown Form	Crown Cover	Leaves	Special Features
125–250	desert scrub	up to 5 m.	cacti dominate landscape	very sparse	spiny small	cacti and other succulents, thorny scrubs
250–500	thorn woodland	up to 10 m.	flat tops	sparse	spiny, sclerophyll hairy, deciduous	cacti, acacia-like trees, thorny plants, grasses
500–1,000	very dry forest	up to 15 m.	large flat tops	few open spaces	many sclerophyll deciduous	less cacti, large acacia-like trees, grasses, very hard woods common
1,000–2,000	dry forest	up to 30 m.	mixed crowns some very large umbrella shaped	no open spaces	markedly deciduous	very large isolated diameters and crowns, few grasses, few buttresses
2,000–4,000	moist forest	up to 50 m.	mixed crowns	no open spaces	evergreen	palms common; on flat areas swamps are widespread; many epiphytes; increase of trees with buttresses
4,000–8,000	wet forest	up to 50 m.	mixed crowns	no open spaces	evergreen	increase in number of palms and swamps; many epiphytes, buttresses common but diameters not very large
above 8,000	'true' rain forest	up to 40 m.	mixed crowns palms dominate	some open spaces	evergreen	abundance of palms, bogs on hills, small standing volume of trees, small diameters

The Holdridge (1947) classification of life zones is used to indicate rainfall boundaries and names of most of the climatic natural vegetation.

East Asian or Temperate Monsoon, found in central and northern China.

Stamp considers that the application of W. Köppen's scheme[125, 126] for classification of climates to Asia is far from satisfactory and that the arbitrary lines separating even the major zones have little meaning. Nor does he consider that C. W. Thornthwaite's scheme[219, 220] will be acceptable to the student of Asian geography, since it produces 'too many strange bedfellows and . . . too many anomalies'. A statement of the position in terms of the location and seasonal movement of Asiatic air masses may be more profitable.[80]

When considered in relation to the growth of plants and animals, a monsoon climate is to be regarded more as a climate of seasons with winds from opposite directions. The mechanism which produces these winds is complex, under the effect of the following influences (E. H. G. Dobby):[65]

(1) A permanent high-pressure belt of anticyclones is associated with the Tropic of Capricorn over Northern India and the Pacific. The Thar Desert is a consequence of this, as are the deserts of west coasts in similar latitudes of North Africa.

(2) A cold-season anticyclone over Central Asia induces outflowing cold, continental air from the 'frozen heart of Asia', descending into Eastern China and there converging with moist, maritime air outblowing from the North Pacific high-pressure belt. The Tibetan and Himalayan relief systems are high and broad enough largely to isolate Central Asian air from India, which at that season develops its own anticyclone system inducing convergences around it.

(3) The Intertropical Front is a phenomenon of the tropical zone, with variations of movement over Monsoon Asia which account for much of the peculiarities of local raininess.

The effects of the air movements associated with the intertropical front are modified within Monsoon Asia by:

(1) Altitude effects causing 'relief rains' and 'rain shadows'; the consequences of the great relief barrier of the Himalaya and the Arakan and of rain shadows in Central Burma and Lower Siam are evident on any rainfall map.

(2) Local 'continental' effects particularly over the Deccan

which tend to induce seasonal wind movements by reason of its own size.

(3) Convergences of sections of maritime air which meet after having taken slightly different routes, during which they have acquired slightly different moisture characteristics; storms due to this re-convergence appear over continental South-east Asia and over the Philippines.

A contrast should be noted between North-western India and Pakistan, and Northern China, margins of Monsoon Asia that resemble one another as zones of dryness and continentality. The former is a self-contained unit, largely isolated from Central Asia, and more subject to interplay between tropical desert and mountain conditions. The latter is dominantly influenced by Siberian air converging with that of the Pacific and by its high latitude.[65] On the Asian mainland and in Japan there are alternate seasons of continental and maritime air, giving rise to an alternation of dry and wet seasons. On the islands this is not so obvious, as all winds pass over considerable stretches of sea and are therefore maritime in nature and moisture content.

In mainland China, all the many variations that occur in the regional climates are variations on one basic theme of causation, the monsoonal rhythm arising from the continentality of the land mass of Asia, according to T. R. Tregear.[221] The variations caused by contrasting ranges of seasonal and diurnal temperature and precipitation and of calm and storminess all reveal in greater or less degree their causal relationship to the rhythmic seasonal change from a dense high-pressure centre over the land in winter to a low-pressure area in summer (see further discussion on China in Chapter VII).

The Dry Tropics

Since, to so much of the transect, the terms 'arid' and 'semi-arid' are loosely applied, it is necessary to be more precise in their definition.[253] A more accurate connotation may be based upon the position of any particular area in relation to a threshold of expected or effective rainfall, a threshold which will vary in relative position from one region to another according to the growing season of plants, to the nature of the climate (summer rainfall, winter rainfall, monsoonal, etc.) and to altitude, latitude and other factors. On one side of such a threshold the conditions are said to be arid and at least marginal if not quite unsuitable for crop cultivation—extensive grazing is the only reliable form of land use. On the more humid side of this threshold, conditions are said to be semi-arid, and crop cultivation using dry-farming techniques and drought-resistant or drought-escaping crops can be recommended. Naturally neither this nor any other threshold on the agroclimatic gradient is precisely fixed along any given isohyet or any lines of expected rainfall. There is in many regions a transition zone in which rainfall and therefore crop cultivation are quite unreliable, but into which agriculture is frequently if not generally forced under a variety of social, demographic and economic pressures.

The geographical location and extent of the lands coming within the individual segments of an agroclimatic gradient vary widely with the seasons, and so also do the actual and potential types of land use. For instance in the kharif or monsoon season in the 'drier' parts of India, much of the country is *subhumid* to *humid*; in the rabi or winter season the greater part of the same land is *semi-arid* (when drought-resistant crops adapted to lower temperatures may still be grown on light rains or residual soil moisture following the growing of a monsoon crop), or *arid* (when crop cultivation is not possible, grazing is the only form of utilization, and domestic livestock wander indiscriminately over both the cultivated stubbles and the natural grazing lands).

Z. Naveh[156] has reviewed the literature relating to the delimitation of the dry tropics in Africa. D. L. Blumenstock[15] has discussed the distribution and characteristics of tropical climates in general and the boundaries between the wet and dry tropics. Although Köppen's thermal-illumination boundary of 18° C for the mean coldest month may be equally applicable to the dry tropics, the actual moisture boundary is more difficult to define because it depends much more upon local soils and topography. Thornthwaite's formula is inaccurate when applied to the drier parts of Africa, and Meigs' map of arid and semi-arid homoclimates ignores large areas in the tropics. J. Phillips[171, 172] has used climax communities as indicators of agricultural potentials in the tropics; all his mild sub-arid and arid wooded savannahs and desert and sub-desert areas and those parts of his subhumid wooded savannahs having severe moisture stress for three to five months or more should be included in the dry tropics. It may be, according to Naveh, that detailed local studies of seasonal rainfall patterns and their effectiveness for actual pasture and crop production (Slatyer)[203] could provide a more reliable guide to a precise delineation between the wet and dry tropics.

S. K. Seth and M. A. Waheed Khan[201] tried to apply Emberger's climatic index to the dry zone of India in order to delimit ecoclimatic regions or zones. As the climatic subtypes so formed did not agree with the types of natural vegetation in the region, the concept of Effective Temperature Index was evolved. This sufficiently modified Emberger's formula to bring it into agreement with the types of natural vegetation. It may be asked whether evapotranspiration, actual or potential, might be used to define agroclimates or ecoclimates more precisely; as, however, actual evaporation data are meagre and transpiration from natural vegetation is difficult to measure, a rational classification of climates on this basis has not yet been evolved for India.[214] In any case, one must accept H. L. Penman's[167] warning about a much debated part of the hydrologic cycle, the transfer of water vapour to the atmosphere by vegetation, a subject characterized by a 'chaos of observations, guesses, inferences, and dogmatic assertions that find their way into print as facts'.

Drought may be rigorous or mild in intensity, long or short in duration, seasonal or aperiodic in occurrence.[144] Its repercussions are equally as varied. An expected period of drought may pass unnoticed, whereas a slight extension over the critical margin of exhaustion of water resources or the occurrence of drought at an unexpected time could adversely affect different ecosystems and the land-use system.

The Humid Tropics

An attempt to delimit the humid tropics has been made[69] for the use of UNESCO in designing its Humid Tropics Research Programme, following a meeting held at Kandy, Ceylon, in March 1956. It was found that any solution which might be reached would be purely arbitrary, the main point at issue being how much seasonality one would allow and still classify an area as humid tropical. Two approaches were adopted. A. W. Küchler prepared a map based on the assumption that a generally mesophytic vegetation showing no great predominance of xeromorphic characteristics or putative adaptations to reduce transpiration, indicates an excess of available moisture over potential evapotranspiration. This excess seemed to be an acceptable criterion of humidity, allowing for local variations in exposure, drainage, wind, cloud, moisture-holding capacity of the soil, etc. B. J. Garnier at the same time prepared a map of the humid tropics based on meteorological data according to a formula of his own.

Change of Climate

From the plant ecological and agricultural points of view, it is probably correct to consider four broad types of alterations in climate which vary in their biological and economic significance:[251]

(1) Major changes, into or out of ice ages or pluvials.
(2) Minor changes which persist for 100 to 300 years.
(3) Variations or trends which are experienced for 10 to 50 years.
(4) Changes induced by the action of man in 'destroying his landscape' which may be distinct from or may be interwoven with types (2) and (3).

Evidence under type (1) may be deduced to study the migration of genera and species to and from Malesia and Australia. The zones of the transect are also highly susceptible to the short-term variations and trends in type (3); above all, they have been subject to the changes induced by man.

The effects of major changes in climate, e.g. those related to the last Pluvial period, are fairly well known or at least relatively to study in terms of geographical movements of the major types of vegetation and to some extent of their component species.[25] It is less easy to find reliable data regarding the disappearance and subsequent return of plant communities and individual plant species during the periods of minor change which persist for 100 to 300 years. This is partly due again to the confusing interference of man-induced devegetation or lowering of the status of plant associations in the ecological succession, and partly to the absence of qualified observers.

The conditions of aridity in western Rajasthan set in, according to P. C. Raheja,[180] in the geologically Recent and Sub-Recent time consequent upon the rise of the Himalayan and Siwalik ranges. A similar review was presented by S. K. Seth to the UNESCO/WMO Symposium on Changes of Climate in 1961.[200] In the protohistoric and prehistoric times, the whole region was well drained by mighty river systems. With the receding of Tethys sea and the gradual drying-out of the river system, desertic conditions set in about 2,000 to 1,500 B.C. This process was hastened by the invasion of Huns and Gurjar nomadic tribes from Central Asia. There is evidence that precipitation is slowly decreasing in this region as the heat and moisture balance is becoming adverse. It is hoped that the present measures directed towards the rehabilitation of the vegetation of western India may have a favourable influence on at least the macroclimate.

With reference to short-term variation, there appear to be no important studies along the line of the transect, but we have the results of detailed studies in a completely different environment, made before, during and after a series of drought years in the Great Plains of the United States of America. These studies were made by Weaver and his associates in the University of Nebraska, and have been thoroughly documented in *The Grasslands of the Great Plains*,[246] and elsewhere.[2, 3, 245.]

The great drought of 1934–40 was so severe and so prolonged that only fragments of the major consociations of which the true prairie is composed remained at its conclusion. Moreover, invading xeric grassland dominants had taken over most of the area and with other species had almost overwhelmed any climax survivors. In 1941–2 a changed environment prevailed. The major area was at that time occupied by thriving invaders, dominants of the much more xeric mixed prairie to the west. If climax prairie was to return, these invaders had to be largely replaced. Two decades were required for this process to be completed.[245] The deterioration during the drought years themselves was accentuated by overgrazing of the drought-suffering vegetation and by smothering of the plants by dust from the cultivated land. Thus we have modern consequences of the misuse of the natural vegetative cover and the extension of cultivation beyond the climatic threshold superimposed upon the normal reactions of the grass species to recurrent drought. These grasslands had certainly suffered from droughts as part of normal climatic fluctuations for centuries, but probably before overgrazing and the ploughing up of the Great Plains they suffered less and recovered more rapidly. The 1953–5 drought has also been recorded in terms of regression and recovery of the plant cover of the Great Plains.

In Professor Weaver's own words: 'Droughts come and go; native vegetation is not entirely destroyed; some grasses lie dormant for 5–7 years and then revive. The grassland-prairie border is fairly stable despite drought, as regards man's lifetime. It is true that over a belt 50–150 miles wide, Great Plains vegetation moved eastward into normal true prairie in 1933–40, as we have amply recorded, but it is also true that the reverse of this occurred following the cessation of drought in 1941–50. Not much change has occurred along grassland-forest borders of which I am aware. Death of shrubs in brushland and savannah in the south-west has occurred to some extent.' There is a vast field of research on these lines awaiting the ecologist and the climatologist working together on the grass covers of the transect.

Mapping of Climatic Regions

It seems to be generally concluded that the classifications of world climates proposed by Köppen and Thornthwaite are open to considerable criticism in those parts of Africa and Asia covered by our transect.

On the global scale, we await the map of monsoonal climates being produced by the World Meteorological Organization. In the meantime, we have other maps, including that prepared by C. Troll and K. H. Paffen of the seasonal climates of the earth.[127] This is the first map to be based on the seasonal sequence of natural phenomena in the different climatic zones of the earth. The regionally different interweaving of insolation, temperature and humidity seasons has many direct ecological consequences. The relationship between seasonal climates and vegetation is indicated in the legend which lists thirty-three subtypes.

As far as Africa in particular is concerned, there is the climatological atlas prepared by the African Climatology Unit (directed by S. P. Jackson), on behalf of CSA/CCTA.[100]

Agroclimatic Zones and Regions

The first FAO/UNESCO/WMO joint project on agroclimatology in Lebanon, Syria, Jordan, Iraq and Iran was concerned with an arid and semi-arid winter-rainfall environment. The area selected for the second survey was the 'tropical area with a semi-arid climate of summer rains in Africa south of the Sahara, on the belt bounded to the north by the arid steppe and desert and to the south by humid savannah and forest', already defined at the beginning of this chapter. J. Cochemé[48] reported to the UNESCO Symposium on Methods in Agroclimatology at Reading in July 1966 (see also Appendix 1).

An investigation was carried out of a belt of territory in West Africa 600 km. long and about 1000 km. wide bounded to the west by the Atlantic Ocean and to the east by the border between the Republics of the Sudan and of Chad. It includes parts of the following countries: Senegal, Mauritania, Mali, Upper Volta, Ghana, Togo, Dahomey, Niger, Nigeria, Cameroon, Chad. It is represented, together with mean annual rainfall, in Fig. 14.

This vast continental expanse is virtually free from maritime influence from the west and shows very little geographical relief. In terms of general circulation, it is situated between the tropical anticyclonic belt and the equatorial convergence, and it is influenced by both in varying degrees during the seasons. It is remarkable for the uniformity of its climatic zones and the steadiness of its climatic gradients along the meridians. Here 'zone' is used to mean not only a surface distinguished by some feature but also a strip lying along the parallels of latitude.

This zonation, in an area where the input of solar energy at the upper surface of the atmosphere is fairly constant, is caused by the annual northwards surge and retreat of the inter-tropical discontinuity which, aligned latitudinally, substitutes every summer over the area a moist and unstable air mass yielding rain for the stable and dry anticyclonic current which covers it the rest of the time. The summer rains are, on average, heaviest in August everywhere, and their duration varies from about six months in the south to two in the north.

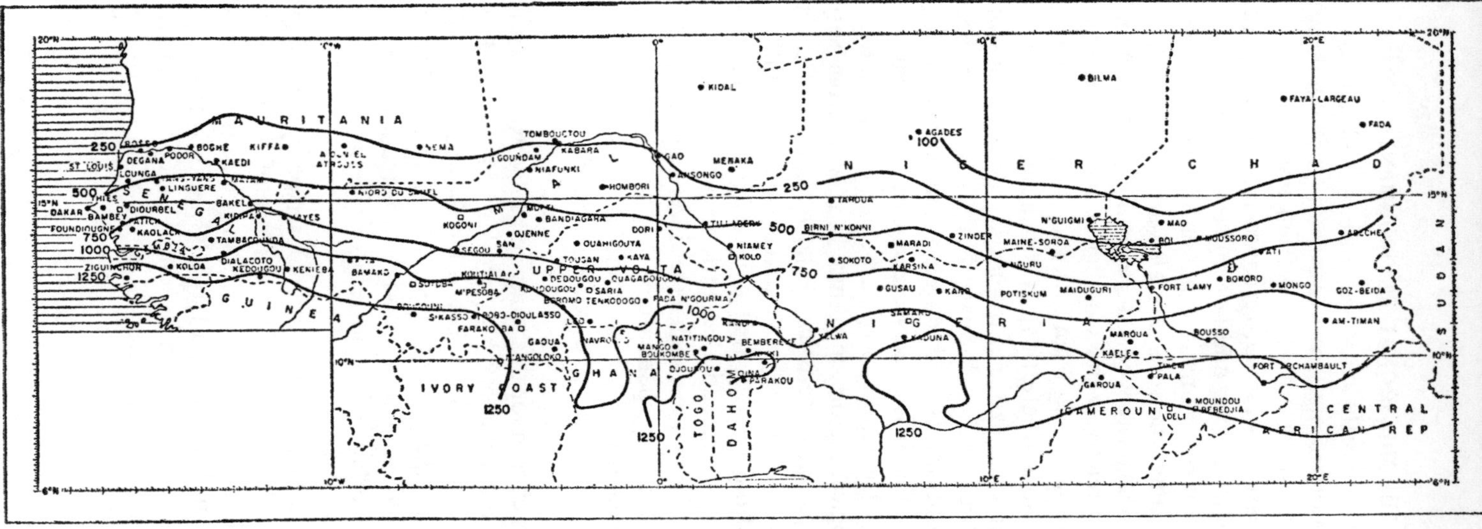

FIGURE 14

FAO/UNESCO/WMO Agroclimatology Survey: Summer rainfall semi-arid area south of the Sahara, showing mean annual isohyets[48] (see also Fig. 55)

Precipitation is mostly due to disturbance lines. These are rows of self-propagating convective clouds giving showers and thunderstorms which are aligned from north to south and move from east to west. Disturbance lines may be several hundred kilometres long and last for several days.

As a result of this climatic zonation the native vegetation and, very broadly, the soils, are also arranged zonally. The vegetation is a sparsely wooded steppe in the north changing southwards to relatively dry woodlands and savannahs.

The soils show, going from north to south, first a narrow zone of grey soils where there is insufficient rain for much leaching, next a main zone of tropical ferruginous soils where the iron has been leached downwards, in increasing degrees as one moves southwards, and, on the southern border, forrallitic soils in which it is mostly the silicates which have been leached away. The remnants of past climates, as well as more strictly geological factors, introduce further heterogeneity.

The principal characteristics of the agriculture of the area surveyed are that it is very extensive, that soil fertility is restored by natural fallows, that cultivation and animal husbandry are separate, and that little or no use is made of intensification techniques, such as fertilizers, plant protection, plant introduction, mechanisation.

These characteristics, largely imposed by the climate, become more pronounced and unavoidable the further north one goes.

Since the area chosen for this study coincides exactly with the western end of our transect, those working in semi-arid monsoonal environments may look forward to the final report of this second survey with the greatest interest. We may also hope that the next FAO/UNESCO/WMO study will be made in the region that is most important of all in terms of human population, namely, monsoon Asia.

The staff of the Main Geophysical Observatory in Leningrad has compiled an Atlas of the Heat Balance of the Earth, published in 1963. The map of total radiation (Fig. 15) presented by M. I. Budyko[23] in his paper to the UNESCO Symposium on Methods in Agroclimatology in Reading in 1966 shows that the distribution of annual quantities of radiation in the tropics is closely connected with the distribution of cloudiness. These data

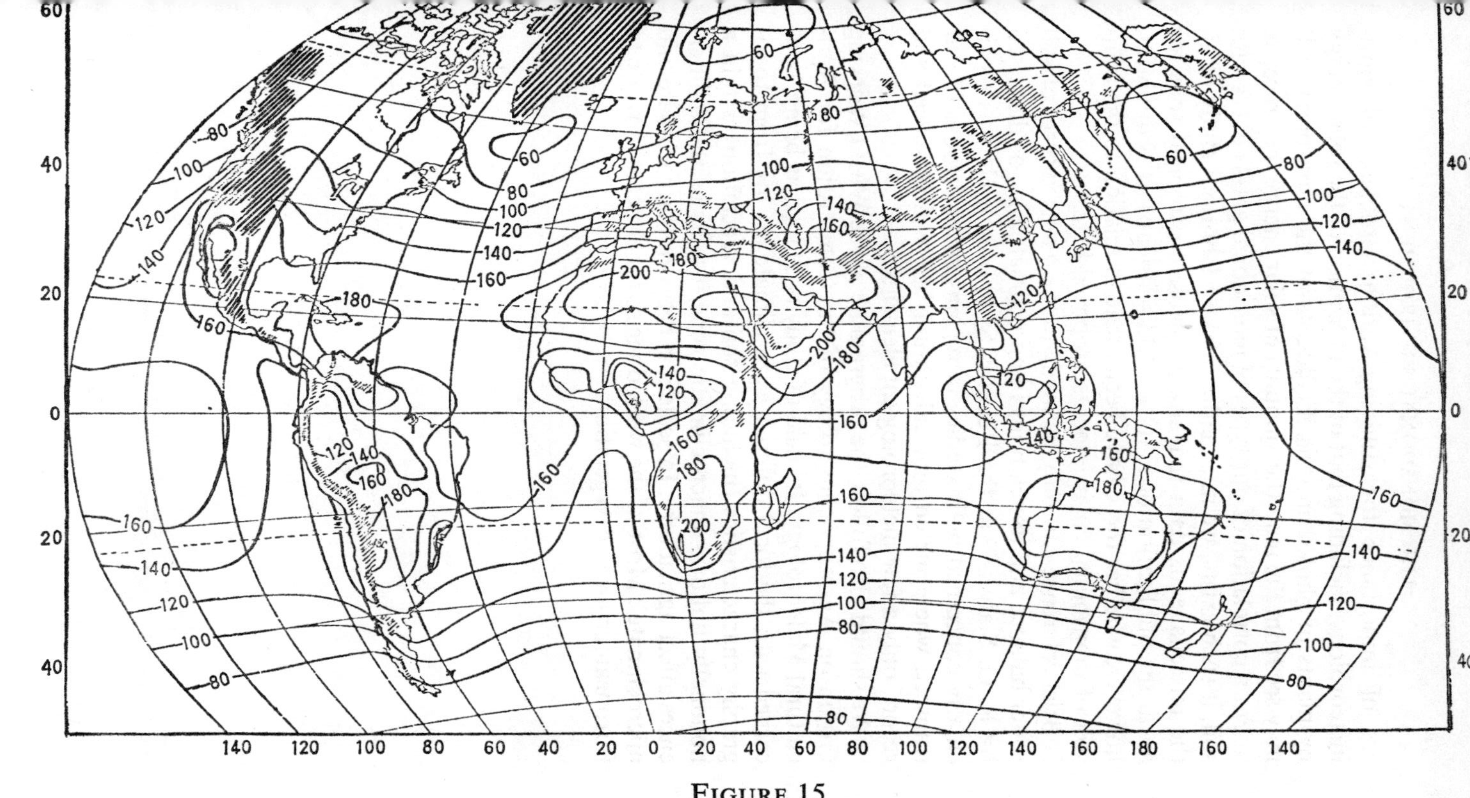

FIGURE 15

Total radiation in kcal/cm²/year, according to the Geophysical Observatory, Leningrad[23]

are of importance for studying various problems of agrometeorology, including that of the utilization of the solar energy by plants in photosynthesis. Ecologists of the monsoonal belt may see from the map the amounts of solar energy they receive.

For a correlation of our types of monsoonal grass covers with soil, if such exists, we must either make a laborious study of all the miscellaneous data that are available on tropical soils, or await the publication of FAO's world soil maps. It is doubtful, however, whether these have been made in relation to the transect of monsoonal environments proposed here.

Brief reference can only be made to the whole subject of soil water balance and its relations to agroclimatology, as discussed by R. O. Slatyer[204] at the Reading Symposium. He was particularly concerned with studies in regions such as those of our transect, where meteorological stations are widely spaced and record only a few meteorological elements.

We should also follow the programme of the WMO Working Group on Agrotopoclimatology of the Commission for Agricultural Meteorology. Its field of study is defined as being 'concerned with the local differences in climate arising from topography characteristics (including soil and vegetation) within a nominally uniform macroclimatic zone insofar as these differences affect agriculture. In scale it is between macroclimate and microclimate'. Most of the work done so far appears to relate to European crops and vegetation.

Chapter III

PLANT BIOLOGY AND ENVIRONMENT

Purists in the fields of plant ecology and of meteorology will no doubt consider it dangerous to propose the existence of a vast transect of Africa, Asia and Australia, merely because that environment seems to provide conditions for the growth and reproduction of a relatively limited number of grass genera and species. This comparability of environment and response does, however, exist, as Chapters IV to XIII show. In this chapter it is proposed to refer very briefly to a number of subjects which would seem to deserve further consideration in any coordinated study of plant behaviour in a monsoonal ecoclimate. We have in this zone a wide range in total annual rainfall, in duration of dry season and in the fundamental aspects of geomorphology, geology and soils. It would appear to be possible to bring the grasses of the monsoonal transect into perhaps four major groups:

(*a*) the arid and semi-arid group, to be found primarily at the western and central parts of the transect, as far as Rajasthan;

(*b*) the medium-rainfall group found mainly in India, Burma and northern Australia;

(*c*) the group of the high-rainfall monsoonal and equatorial or ever-wet conditions in south-east Asia, including Malesia;

(*d*) the group of genera and species adapted to a hot monsoonal summer and a cool to cold winter, such as obtain in China, Japan and Taiwan, particularly at higher altitudes.

Scattered throughout the eastern parts of the transect we have larger areas or smaller discontinuous islands of high-altitude grass communities, comprising genera of temperate latitudes (Himalaya, Taiwan, Malesia, Papua and New Guinea). It should be the objective of research to determine how these different conditions govern the growth period and growth habit of grasses, the nature of the communities of plants in which these grasses find themselves, and therefore ultimately the types of animal husbandry that may be achieved or introduced.

Geomorphology

This subject may be considered either on the macro- or local scale. In the first category may be mentioned the structural pattern of Asia, its land forms and physical patterns, described by E. H. G. Dobby[65] in Chapter 2 of his *Monsoon Asia*, and shown in Fig. 16. This pattern governs the extent and nature of the monsoonal climate to which reference is made in Chapter II, and through the climate the natural vegetation and types of grass cover.

On the more local scale, it is generally agreed that ecological systems are clearly associated with geomorphological systems, although the closeness of the relation may be masked by the

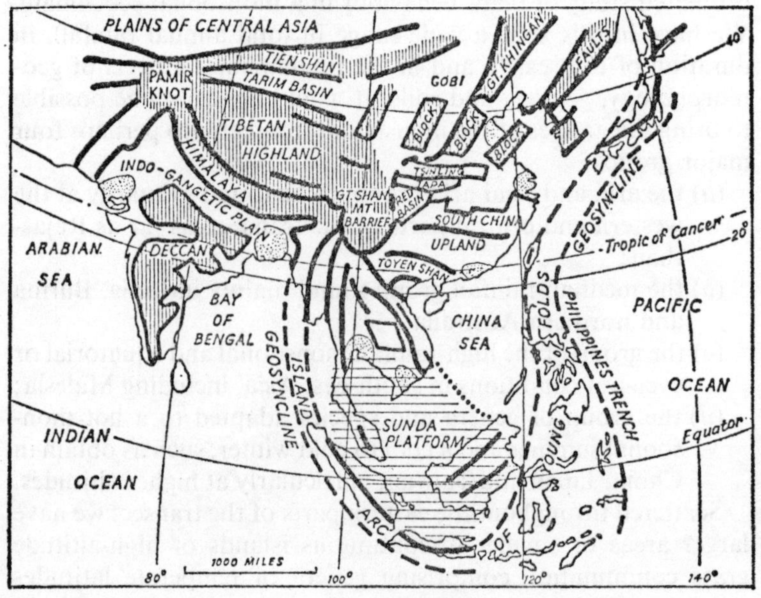

FIGURE 16
The structural pattern of Asia, according to E. H. G. Dobby

interference of the biotic factors of grazing, burning, cutting and the spasmodic clearing of vegetation for shifting cultivation.

Within the transect the most important work in this field is, of course, that of CSIRO in Northern Australia. The studies of Ch. Rossetti,[191] to which extensive reference is made in the section on West Africa, have a strong geomorphological bias. M. Kassas has observed the relations between land form and plant cover in the Egyptian desert,[110, 111] and in the Omdurman Desert in the Sudan.[106] In the latter study he concluded that a close relation exists between the plant cover and the land form, the physiographic features apparently controlling the plant cover through its control of the water resources. Kassas's general comments are as follows:

> Datum land, which receives an amount of water equal to the recorded rainfall, does not exist. The differences in surface level, imperceptible as they may be, will cause water to run off from one area and collect in another. The amount of water received depends on the catchment area of the receiving basin: all controlled by topography. The amount of water retained will depend on the texture and depth of the surface deposits (soil). That the soil texture has a controlling effect on the water relationships is best expressed by J. Smith,[205] '. . . the tree species which requires 3x inches of rain on clay soils requires 2x inches of rain on sands'. This has also been suggested earlier when we quoted his definition of the southern boundaries of the *Acacia* Desert Scrub as: 400 mm. isohyet on clays and 250 mm. isohyet on sands.
>
> The sheets of sand drift accumulating on top of sterile gravel desert provide material for studying the relation between soil depth and plant cover. A shallow layer will become completely dry shortly after the rainy season has come to an end. Only ephemerals can survive. A deeper bed of sand will provide room for water storage in the subsoil and will allow for the growth of perennials. On these sand sheets one can follow the gradation from an ephemeral plant cover, to a drought-deciduous grassland, to an evergreen grassland and finally an open scrubland of *Acacia*.
>
> Human interference ranks second only to the water factor in affecting the plant growth. A few kilometres to the west of Omdurman the Forestry Department established a 5 acre

enclosure a few years ago. This has been extended this year (1955). The difference between the plant cover inside the enclosure (20–80 per cent grass cover) and that outside (less than 5 per cent cover) is outstanding even in the newly established enclosure. The enclosure not only protected the plant cover but also allowed for the accumulation of soft deposits (wind- and water-borne) and the building up of the ground level inside the enclosure. This is especially marked along the perimeter.[106]

Turning farther south in Africa, we have Monica M. Cole[49] who believes that the physical environment is of great importance in governing the distribution of vegetation types and plant communities, and who considers that writers on African vegetation fail to explain satisfactorily either the present distribution of forest and savannah or that of the many vegetation associations included within them. These writers have not acknowledged the ever-changing composition of vegetation, the constant struggle of individual species to extend their range and the way in which slight changes in the physical environment may favour some species at the expense of others. They have failed to appreciate that geomorphological processes, themselves governed partly by the prevailing climate, are continuously modifying the relief and drainage, the soils and the micro-climate, thereby creating conditions more favourable for some plants and less favourable for others, and bringing about the extension of some vegetation associations and the recession of others of which only relicts may remain.

Within the climatic limits the composition and character of the vegetation is governed largely by soil and drainage conditions, which in turn are related to geomorphology. Thus, undulating areas fashioned by current dissection are generally well drained and have mature soils able to support forests, whereas level areas subject to deposition or planed by erosion generally have impeded drainage and soils subject to alternating waterlogging and moisture deficiency which support only grassland. The picture is constantly changing. Changes in the composition of the vegetation are, of course, governed by the species present in the area. Here the age of the floral elements may be important, the tendency being for the older elements to persist on the older surfaces where the soils are exhausted and the drainage particularly poor. . . . Throughout the vicissitudes of land sculpture the vegetation has changed as the drainage, soil and climate conditions have been modified by the interplay of pediplanation and dissection and by the stripping of one geological formation and the exposure of another. As one would expect, the valleys and the escarpments have provided the avenues along which vegetational changes have progressed.[49]

Studies of the vegetation mosaic demonstrate the close relationship between the distribution of vegetation and the geomorphology.

It appears that, in India, it is only in Rajasthan that any serious attempt been made to distinguish land-form regions in general, and to make an intensive study of one part of the region in particular. Three primary and three secondary land-form regions in western Rajasthan may be recognized:[11]

Primary

The predominantly sand-covered Thar;
Plains with hills including the central dune-free country;
Hills.

Secondary

Plains with sand dunes;
Plains of predominantly older alluvium;
Deltaic plains.

Within the Central Luni Basin, where the Central Arid Zone Research Institute is working, the major land-forms are:[196]

Plains formed by older alluvium;
Plains formed by younger alluvium;
Blown sands and sand dunes of Pleistocene and recent times;
Hills and mountains of granites and rhyolites of Pre-Cambrian age.

Wind-blown sands and alluvium cover the major portion of the tract concealing the lithology. Geological formations later than the Lower Vindhyans are not found, while Pre-Cambrian formations cover hardly 7 per cent of the area.

According to the 1962/63 Report of the Central Arid Zone Research Institute, the Central Luni Basin, occupied by extensive flood plains, alluvial plains and residual hills, may be divided into seven geomorphic provinces and units that form the basis of land systems and land units respectively. In their evolution the present land forms owe their origin to a wetter climate in the recent past when the region had a more organized drainage system which has gradually died out. The old and disorganized stream channels regulate the subterranean flow of water and these are the potential sources of sub-surface water for stock and irrigation.

In south-east Asia[64] we find the three basic physical units:

(*a*) the Sunda Platform, underlying Borneo, Eastern Sumatra, Northern Java and Malaysia;

(*b*) the Sahul Shelf, to the east of the East Indian archipelago, the common physiographic basis of Australia and New Guinea;

(*c*) a 'ground swell' of young mountain areas fringing and lying between the two other units.

Finally, there are the most comprehensive studies of all made along this transect, represented by the work of the Division of Land Research, CSIRO, Australia; to the reports on Papua and New Guinea and northern Australia, and to papers by C. S. Christian[39, 40, 46, 47] reference should certainly be made.

There is a need for the extension of this work to other parts of the transect. In defining more precisely the relation, if any, between geomorphology and vegetation in general or type of grass cover in particular, some reconciliation between contrasting views seems to be needed, as between the plant ecologist and the geomorphologist. The plant ecologist may stress biotic factors, fire and other direct and indirect effects of human interference, which may be more variable in distribution and intensity. Savannah vegetation has evidently occupied some surfaces for a very long time, but on others natural forest has been converted to savannah by burning and intensified grazing and cutting for fuel in historic times. The geomorphologist accepts the fact that plant communities in existence now do not necessarily represent undisturbed vegetation, the ecological behaviour of which is determined entirely by site.

Within the same land form it is possible from an established vegetation type (tree/shrub/grass communities), presumably governed by geomorphology, to produce quite distinct types of vegetation. Their physiognomy and botanical composition will differ, depending upon type of treatment by man (fuel cutting of secondary forest growth with or without removal of roots; introduction of grazing and browsing animals, cattle and/or sheep and/or goats, each with their special grazing behaviours and preferences; accidental or deliberate burning; and cutting of grass for hay with or without subsequent grazing; shifting cultivation). These questions are of fundamental importance in the study of grass covers and their place in the total vegetation picture.

Geological History

Questions relating to continental drift, mesozoic continents and the migrations of the angiosperms have been reviewed by R. Melville[146] of the Royal Botanic Gardens, Kew. (See also Chapter 15 of *The Geography of the Flowering Plants* by R. Good.[76])

> The origin of the flowering plants, how they came to appear relatively suddenly in the northern hemisphere in the Cretaceous, and how they attained their present distribution are different aspects of one of the outstanding problems of biology. In recent decades it has become increasingly evident that the angiosperms must have had a long evolutionary history before the Cretaceous. . . . The problem is too complex to be solved by any single line of evidence, but requires the correlation of data from many fields including taxonomy, phylogeny, palaeobotany, palaeography, geomorphology, and palaeomagnetism. The critical period for this study extends through the Jurassic and Cretaceous, when Mesozoic land masses were breaking up before their rearrangement into the present continental system. . . . The next phase in the description of Gondwanaland was the development of the mid-Indian rise which continued south eastwards through the junction between Australia and Antarctica. Evidence to date the northward drift of India is scanty, but it was . . . relatively rapid and, in its passage north, the resulting climatic changes caused the almost complete annihilation of its rich gymnosperm flora. Perhaps for the same reason, India appears to have made a smaller contribution to modern floras than its size would warrant.

The relation between geological history and plant distribution comes up again in the geographical chapters on India, Tibet, Malesia and Australia. In the history of India, Tibet and neighbouring regions, the Tethys Sea appears to have played an important role. R. Good[76] states (see also Chapter VI):

> It is generally believed that for a very long period of geological time, perhaps from the Carboniferous to the Middle Tertiary, what is now Eurasia was separated from other land masses on the south by a sea called the Tethys, which extended all the way from the Atlantic to Malaysia. The western part of this sea is now largely covered by the North-African-Arabian deserts, and the Mediterranean is all that remains of it here. The presence of this long east-west sea must have had profound effects on the climate of these parts of the Old World. It has been suggested that warm currents, passing through in this direction, may have modified the climate of what is now Europe sufficiently to allow tropical plants to grow there, or at least much nearer than they do now.

To explain the occurrence in Malesia of genera and species of legumes and grasses now dispersed over remote stepping-stones separated by distances too wide for bridging by natural means of dispersal, C. G. G. J. van Steenis (see Chapter XI) introduces the influence of the Pleistocene Ice Age. This was accompanied by a general lowering of the level of the sea, and the increase of the land masses changed the wind regime and air humidity in Malesia. With the Post-Glacial rise of sea level the present situation was established, leaving a wider spacing between the stepping-stones caused partly by the Post-Glacial increase of humidity, but also by the submersion of large areas of land. The exchange of drought-resistant plants between Asia and Australia is, according to van Steenis, of rather recent occurrence in geological history. (See also Chapter XIII.) The Indo-Malesian element has been isolated in Australia since early Tertiary times.[208]

Soils

While we await publication of the FAO/UNESCO World Soil Map, we may with reference to *Monsoon Asia* only give the soil types quoted by E. G. H. Dobby.[65] The dense populations and the most intensive forms of agriculture of these lands have developed on comparatively new alluvials, the result of large-scale erosion following devegetation of the Himalaya and the hilly country adjoining the alluvial plains of China. The soil types of Monsoon Asia are:

(*a*) The recent alluvials, especially in the deltaic plains, in the Lower Ganges and the North China Plain.

(*b*) The highly varied soils of the mountain and valley complexes of the Himalaya,[254] Eastern Tibet, Inner Borneo, Sumatra and Java.

(*c*) Laterites (the most widespread), evolved under sustained high temperatures and heavy rainfall, low in organic matter, and of limited potential.

(*d*) Desert soils of the Thar Desert and on the boundary of the transect in north-west China.

(*e*) Chestnut brown soils, developed under an original grass or bushy cover, between the desert and the lateritic soils in northern India and between the desert and evergreen coniferous forest in north-west China; developed on riverine deposits in India, on loess in China.

(*f*) Black-earth soils or chernozems in Deccan and south-west Manchuria.

(*g*) Grey-brown podsols in parts of China, Korea and northern Japan.

(*h*) True podsols, most of Manchurian plains.

Plant dispersal and migration

No study of a theoretical global transect would be complete without at least brief reference to the subject in which the disciplines of geological, climatic and botanical history are combined with plant taxonomy and cytogenetics to explain the present occurrence and distribution of plant species. To supplement the information given in the chapters on India, Tibet, Malesia (C. G. G. J. van Steenis), and Australia, one may begin by referring to J. Hutchinson's maps[97] of world distribution, and then, with reference to Africa in particular, to the papers by R. E. Moreau[151, 152] on African biomes in the late Pleistocene, and on certain biological problems in Africa since the Mesozoic.

Theodore Monod has said that the question is of great interest but asks how any definite answer could be given, except perhaps in the case of quite recent introductions historically documented;

how may we distinguish between species that may have travelled from east to west by natural means of dispersal and those that have been dispersed by human aid? For wild species, our present knowledge does not permit any useful conclusions. Such problems are already complex for cultivated taxa, and wild species are not likely to be any easier. Regionally they may be some transport of zoochoric diaspores, for example of *Cenchrus biflorus* and species of *Neurada* and *Tribulus*, but in any direction. The situation may become clearer with the publication of more floras (Uppsala meeting of AETFAT in September 1966).

Highly relevant to the question whether the African flora originated on the spot or immigrated from elsewhere was the theme of the 5th Plenary Meeting of AETFAT (Association pour l'Etude Taxonomique de la Flore d'Afrique tropicale) on the relation of the flora of Africa with that of other countries. J. P. M. Brenan[20] and F. N. Hepper[89] of the Royal Botanic Gardens, Kew, presented papers to that meeting. It appears that we must wait for someone to study the Gramineae from this aspect. Like C. G. G. J. van Steenis, Brenan chose the Leguminosae for his study, and took as his basis, with certain reservations, the assumption that the area where the maximum number of species occurs is its centre of origin.

Of the 229 genera of Leguminosae occurring in tropical Africa, 96 (41·9 per cent) are endemic, and 133 (58·1 per cent) are non-endemic. Of the latter, only four genera are restricted to Africa and India, whereas no less than 31 genera show a distribution: Africa/Malesia/Oceania/Australia. According to Brenan, migration in both directions is clearly indicated. There is a greater similarity at the specific level between Africa and Asia among the dry country genera than the forest ones, indicating that the connection between the forest areas was severed before the connection between the drier areas. Australia cannot be considered as a source of tropical African genera, except in so far as an Indo-Malesian element occurs in both Australia and Africa (cf. Tables 28 and 29 for global distribution of the grasses of Arnhem Land).

In Table 3, F. N. Hepper[89] has shown that over 80 per cent of the species (non-gramineous) taken in his sample analysis occur nowhere else outside tropical Africa. The endemic species can be subdivided into groups according to the vegetational zone that each occupies. Most of the distribution patterns mentioned by E. Milne-Redhead[148] are relevant to the West African flora. Hepper finds the link between west tropical Africa and India and Malesia to be very marked.

TABLE 3

Sample analysis of the distribution of 2,000 species as an indicator for the whole West African flora[89]

	Approx. percentage
West African endemic species	37
Species reaching E. Cameroon to Congo Basin	27
Species reaching Ubangi-Shari to Somalia (± Sahelo–Sudanian domain of Lebrun)	2·5
Species reaching E. Africa (particularly montane)	2·5
Widespread African species (including 1·2 per cent reaching Madagascar)	17
Species reaching tropical and Central America	2
Species reaching Mediterranean region and Europe	0·4
Species occurring in the Middle East and India (including Saharo–Sindian element)	1·7
Species reaching Malesia	2
Species reaching Australia	0·2
Palaeotropical species	2
Pantropical species	5

Dispersal by human agency may be either deliberate, which involves the specialized subject of plant exploration and introduction, or accidental, in a number of ways which may be grouped as follows:[76]

(*a*) by adhesion to moving objects, such as man or moving vehicles or camel caravans;
(*b*) along with the seeds of crops, in which case grasses would be mostly annuals or short-lived perennials;
(*c*) in fodder and packing material, in which case grasses would be mostly perennials;
(*d*) along with minerals such as soil, ballast or road metal;
(*e*) by carriage of seed for purposes other than planting.

In reviewing the gramineous flora of Africa, south-western Asia and India, for example, one has to consider whether the plant communities at any particular site are really indigenous, or whether perhaps some of the genera and species reached that site through the medium of trade and human migration. One thinks of the invasions out of Arabia and along the Mediterranean north coast of Africa—did any of the grass species now fully established there arrive from Arabia at that time, as weed seeds in sacks of cereals, or as escapes from horse fodder and bedding? Did any of our grass species of the Sahelian zone, so easy for east–west travel, reach their present location in this way, or did the Bantu carry southwards from Central to South Africa, through the high country, seeds of any of the grass species that are now found there? Pastoral peoples would perhaps carry mostly perennials, cultivators would tend to distribute annuals. *Colocasia esculenta* came to West Africa from tropical Asia through Egypt before the arrival of the Europeans; plantains came by the same route in ancient times, bananas probably came with Portuguese missionaries; limes introduced from the east by Mohammedans (Okiy).[160] R. Schnell[198] indicates the source of origin of exotic crop plants into former French West Africa, through the Ghana Empire and the periods of Arab and Portuguese domination, but without stating the exact routes by which the introduced plants reached West Africa. A. Chevalier[35] and R. Portères[177] have used linguistic data in studying the origin and possible routes of dispersal of cultivated plants. There has thus been since ancient times ample opportunity for dispersal of grass seeds as weeds in samples of cultivated crops, along the latitudinal zones south of the Sahara. Trans-Saharan migration would be less likely since this would involve the transfer of winter-rainfall species into a monsoonal environment.

In A.D. 45, Hippalus, a Roman sailor, sailed down the Red Sea through Bab el Mandeb to be the first 'westerner' to discover the regularity of the *mausin* winds. This knowledge soon spread and books written in Greek and Latin told of the Indian Ocean and its ports. One guidebook to the Indian Ocean was the *Periplus* (or circumnavigation) of the Erythrean Sea (Indian Ocean). Many and diverse cargoes of plant and other origin were carried in the Arab dhows as they followed the seasonal *mausin* winds. This may have been an important factor in dispersal of grasses between comparable habitats.

Before these traders visited the coast of East Africa, there were invasions or migrations southwards down the Nile Valley from Asia of people speaking Hamitic tongues, through Abyssinia into the land of the negroes in Central Africa. These

Hamites brought the goat, the dog and certain grains, no doubt with a good percentage of weed seeds, all until then unknown in East Africa.

We have seen in Fig. 11 how closely the distribution of the desert locust in Africa and the Near East is related to our monsoonal transect. This impels one to ask whether these followers of overgrazing and cultivation might not also be agents of plant dispersal, and to investigate their eating habits, food preferences and stomach contents.[61] Very dissimilar effects of different food plants on survival and development of locusts have been noted, and indicate a need for fundamental investigations.[227] The rearing of *Locusta* m. *migratoria* on a wide range of plants led a Russian worker to suggest the following grouping of food plants according to their effects on the development of this species.

(1) Plants on which the complete life-cycle can be accomplished—Gramineae and Cyperaceae.

(2) Plants on which hopper development is completed, although with considerable or high mortality, but adults do not mature sexually—Compositae, Cruciferae, some Gramineae (e.g. *Avena*), Plantaginaceae, Leguminosae and Urticaceae.

(3) Plants on which hoppers can reach the last instar, but hopper mortality is considerable at the final moult, or immediately after—Rosaceae, Saxifragaceae, Caryophyllaceae.

(4) Plants on which only early instars survive—Typhaceae, Ranunculaceae.

(5) Plants eaten only by first instar hoppers, which fail to moult—Labiatae, Onagraceae, Chenopodiaceae, Liliaceae, Geraniaceae, Ericaceae, Betulaceae.

(6) Plants not eaten by first instar hoppers, even when starved—Primulaceae, Polygonaceae, Convolvulaceae, Rubiaceae, Ulmaceae, Caprifoliaceae, Salicaceae.

Gramineae also feature largely in the field observations on the behaviour of hoppers of the red locust made by R. F. Chapman,[32, 33] but as this species appears to confine its diet to leaves and flowers of *Cyperus*, *Cynodon* and *Echinochloa*, its role in the dispersal of grasses would appear to be doubtful.

Life Forms

We may first quote from S. A. Cain.[27]

After a strong interest in the life-form problem around the turn of the century when ecology was developing as a recognized science, there has been little sustained interest in life-forms . . . and no system other than Raunkiaer's[184] has been used widely. . . . American ecologists have been little interested in detailed studies of community composition and structure, and have been more interested in succession and studies of the factoral composition of the environment. There has also been some adverse reaction to life-form study because of physiological investigations which have shown that a plant's functioning is not always what it would seem to be on the basis of structure—that structure and function are not uniformly one—and that adaptation may not reveal itself on the surface. The first duty of the ecologist, however, is to the plant and the plant community; and the life-form, whatever its adaptational significance, is an important part of vegetational description ranking next, perhaps, to the floristic composition of the community.

Although life-form spectra have been employed for general climatic analysis, there are enough studies available to suggest that they may be even more useful for the analysis of climatic variants and microclimates. Also it seems probable that life-form spectra when based on associations, not in the large sense of the association (cf. Clements) but more or less narrowly conceived, can reflect habitat changes with succession, important edaphic conditions, etc. The usefulness of life-form spectra for such investigations would seem to depend largely on some means of evaluating the species composing the flora of the communities . . . life-form spectra for climax or near-climax vegetation are the only ones that can reflect sensitively climatic correlations. Spectra for association variants might reflect minor climatic variations. Spectra for successional communities of various sorts might reflect edaphic conditions and give a good measure of the changing environment as succession proceeds. These, then, are some of the lines for future study that should be profitable.

The grasses (and other species) of Rajasthan have been classified into four habitat forms in relation to their capacity to withstand long periods of drought:[202]

(*a*) Perennial drought-resisting plants:
Extremely xerophytic tussock grasses—
Panicum turgidum *Cymbopogon jwarancusa*

(*b*) Perennial drought-evading plants:
Dichanthium annulatum *Bothriochloa pertusa*
Lasiurus sindicus *Eleusine compressa*
Sehima nervosum *Dactyloctenium sindicum*
Eremopogon foveolatus *Cenchrus* species
Cymbopogon martinii

(*c*) Ephemeral drought-evading plants:
Oropetium thomaeum *Enneapogon elegans*
Latipes senegalensis *Elyonurus royleanus*
Dignathia hirtella

(*d*) Sand binders and sand dwellers:
Saccharum spontaneum *Saccharum munja*
Panicum turgidum *Cenchrus* spp.
P. antidotale *Lasiurus sindicus*
Lasiurus hirsutus *Eleusine compressa*
Cymbopogon jwarancusa *Dactyloctenium sindicum*

A similar classification has been used by R. A. Perry[169] in northern Australia.

In grasses, the relation between the seasonal climates and tillering and reproduction is very important. M. Lazarides, M. J. T. Norman and R. A. Perry[130] working at Katherine, Northern Territory, Australia, studied these characters in six perennial and two annual grasses. There is a 5-month growing season. Most species attained maximum tiller development early in the wet season, but *Sorghum plumosum* produced a post-reproductive flush of vegetative tillers, which remained alive during the dry season. Burning severely depressed growth of *Themeda australis*, but stimulated tillering of *Heteropogon contortus*. *Chrysopogon fallax* was unusual in spreading by rooting stolons as well as from seed. Preliminary indications are that dry matter, nitrogen and phosphorus are concentrated in the bottom few centimetres of plants of these four grasses, even though the plants reached a height of 1·5 to 2 m.

Taxonomy and Geography

N. L. Bor[19] has stated:

> . . . in the Gramineae there exists the most extraordinary mosaic of characters which even now are changing in response to the forces of evolution and selection. Indeed, the mosaic of simple and complex characters in the same plant or even in the same spikelet would seem to indicate that the evolutionary changes of the past were not always at the same pace or even in the same direction . . . we are at the moment witnessing a revolution in ideas concerning the taxonomy of the Gramineae. New concepts, new methods of approach, all contribute to the elucidation of this most difficult family, but it seems obvious that a great deal more information must be gained before even a tentative scheme with a moderate chance of acceptance can be produced.

What contribution can the grass geographer make to this progress? Field workers with large tracts of country at their disposal may be able to follow a grass species from its centre of maximum frequency along the climatic gradient to the points at which it becomes rare or disappears. On a wider scale of time and space, such geographical studies might begin on the basis of the theory by J. W. Bews.[10] He postulated that the progenitor of grasses arose in the humid tropical forests, and that migration to less uniform, more exacting and variable environments was accomplished by mutation and natural selection. Modifications of various kinds in the primitive progenitor[95] would ensure greater protection for the seed and a complicated dispersal mechanism as the present-day end-products of trends of evolutionary change.[19] But who is to know in which direction a species may be progressing through mutation and natural plant selection, outwards towards the arid and semi-arid unknown, or back to the comforting habitat of the humid tropical forests?[270]

The logical outcome of this hypothesis, against which a number of contrary examples can be produced, is that taxonomic advancement proceeds in the opposite direction to ecological progression. For example, the species of *Aristida* in India, or of *Stipa* in Iran and elsewhere, have become the most advanced taxonomically in their progress into extreme environments. At the same time and for the same reason, they may be regarded as the lowest in ecological succession on any particular site on which they occur, increasing in frequency with deterioration and particularly with increasing aridity of the environment due to the action of the biotic factor in one of its forms. From his long experience in Iran and other parts of the Near East, however, H. Pabot doubts whether it would be safe to generalize in this way with regard to the successional status of species of *Stipa*. It is possible to do this only on the basis of a reconstruction of a theoretical succession in a region where excessive grazing, cutting and other forms of misuse have reduced the vegetative cover to a very low ecological level.

Grasses that pioneer into semi-arid and arid environments must possess a great reserve of genetical variability—the flexibility of species becomes important. Different specific spectra may arise in wet and dry years in the arid zone with wide variability of rainfall between seasons. Types will therefore always be available within a natural population to cope with conditions in whatever ecological niche they may find themselves. The lower the rainfall, the wider the annual variations in its amount and geographical distribution. Short-term and medium-term climatic changes may also have to be overcome.[251]

The major role of polyploidy in the evolution of grasses has been in the fixing and spreading of hybrid combinations at either the varietal, subspecific or specific level.[141] Polyploidy has also provided one of the most rapid known methods of producing radically different but nevertheless vigorous and well-adapted genotypes. Polyploids in general display wider ranges of tolerance of extreme climatic and edaphic conditions; their spread is favoured by the availability of new ecological niches in rapidly changing environments (due to climatic fluctuation and change, or to increasing aridity following devegetation).

Auto-polyploidy and apomixis may together facilitate the advance of species into adverse environments. To what extent is apomixis (a term covering all types of reproduction which replace or substitute for the usual sexual process) an important characteristic for the spread of grass species in arid zones,[81] with particular reference to the revival of variability through the crossing of apomictic lines? Finally, is it true that grasses of the arid zone provide a higher degree of variability available to the plant breeder than the types of the humid tropical forest with its uniform and unchanging environment?[270]

But genecology may be said to have been so far too heavily autecological; it is time to progress to the genecological consideration of synecological situations, and particularly to the response at the association level to environmental changes. Erna Bennett[9] has stated:

> As far as the individual species component of the association is concerned, not only is the physical environment more or less randomly fluctuating, and selection acting in favour of flexibility, but also the biotic environment is changing, and the extent to which it does so randomly will depend upon whether and how much the individuals of its component species are able to respond flexibly. Survival and/or stability of the association would be likely to depend upon the amplitude of the fluctuations on the one hand and the tolerance of its members on the other. Though in general the latter will be at a selective advantage, the evidence of fluctuating vegetative cover seems to suggest that the possibilities of flexible responses have definite limits—i.e. there are definite limits to (or limits to the possible range of) homeostasis or plasticity. With crop plants we are more interested in homeostasis than in plasticity. In any case the whole association is a sort of feed-back system. Ecological analysis might be possible if some expression could be found for general community response based upon the adjustment patterns of the species composing it.

Annual and perennial habit

It should be considered whether all the types of grass cover in our transect are composed predominantly of perennial species. The site data in *The Grass Cover of India* show that the percentage of annual species increases with greater intensity of the biotic factor of grazing. But are there conditions where the grass community is of a purely annual type which will not progress towards a perennial type no matter how effective and long in duration the protection against biotic factors may be?

The position seems to be somewhat similar in India and in the Sahelian zone of Africa south of the Sahara. In the really arid areas along the desert fringe, one may in good years find a certain proportion of annual grasses among the dominant perennials, but in bad years, it is the perennials that persist. Under higher rainfalls, the annuals may dominate, and may under certain conditions of soil, slope and ecoclimate remain as the only types in the grass cover. W. Zeller has reported on this question from observations in the northern part of Saudi Arabia, about 12 km. south of Saqf, lat. 26° 50′ N, long. 41° 05′ E. On gently inclined gravel surfaces, 1 to 3° per cent slope, the distribution of vegetation is shown in Fig. 17.

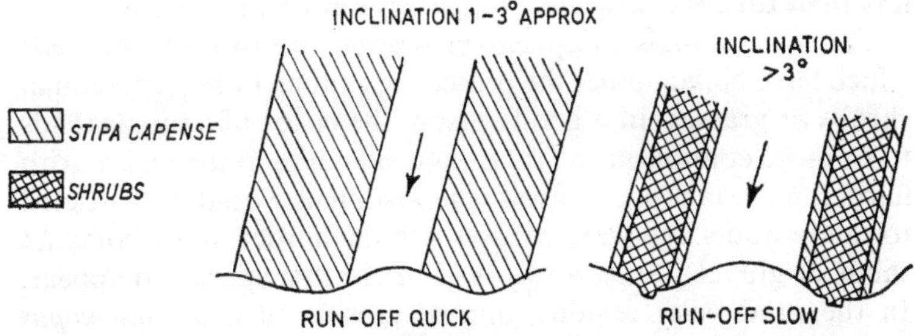

FIGURE 17

Conditions for dominance of annual grass or perennial shrubs in Arabia, according to W. Zeller

In the first case, if the slope is slight, run-off is slow and surface erosion by water negligible. The surface is therefore not furrowed. It is in many cases gently undulating. The vegetative cover is mostly annual, chiefly *Stipa capensis*, which colonizes more particularly the depressed portions of the land. The strips of vegetation are often very broad.

If the slope is more than 3°, run-off is rapid and surface erosion cuts the soil surface, and the vegetation is predominantly perennial—*Lycium arabicum*, *Haloxylon salicornicum* and sometimes *Panicum turgidum*. The strips of vegetation are mostly narrow and confined to the runnels.

These two examples occur on surfaces that are adjacent and therefore presumably exposed to the same intensity of grazing.

It may be assumed that factors other than grazing determine the existence of the annual and perennial types of vegetation. It seems to be a question of water supply, which appears to be critical on dense gravelly surfaces due to considerable surface drainage. Under certain combinations of surface and slope, annual vegetation is the ultimate stage of evolution or climax.

Most of the annual species in the plant communities in Kuwait[117] grow during the cool season—November to April. Although widespread germination is dependent upon rains in late October or early November, the main growth period can be expected in February/March when air temperatures are above 25° C and providing that at least 60 to 70 mm. rainfall has occurred in December or January or before. When rainfall is less than this, the strips of annual vegetation are narrow.

The first annuals to appear in Kuwait are those from seeds which have blown under the protective canopy of large perennial shrubs or grasses. In a good season, the cover of annuals tends to be more continuous on sands and silts than in the region with harder, more compact gravels where annuals would be restricted to sandy wadis and watercourses. On the harder, more compact sands or gravels, *Stipa tortilis* will be the first species to appear. In the sandy-silty regions, annual grasses such as *Sphenopus divaricatus*, *Koeleria phleoides*, *Cutandia memphitica*, *Hordeum glaucum*, *Schismus barbatus*, *Bromus tectorum* are common; the annual legumes, *Medicago aschersoniana* and *Lotus pusillus* are widely distributed, together with, in sandy areas, *Hippocrepis bicontorta*, *H. unisiliquosa*, *Ononis serrata*, *Trigonella stellata* and *T. hamosa*. *Plantago albicans* is widely distributed, together with *Emex spinosa*, *Brassica tournefortii*, *Schimpera arabica*, *Neurada procumbens*, *Paronycha arabica*, *Erodium ciconium*, *E. deserti*, *Launaea nudicaulis* and *L. arabica*.

The growth of annuals is directly governed by the amount of rain; with 100 mm. of annual rainfall, upright species reach a height of about 15·2 cm. while prostrate species have a spread of 10·2 cm. The maximum vertical penetration of roots is 7·5 to 10 cm., but *Launaea nudicaulis* reaches 20 cm. The lateral rooting of plants such as *Medicago aschersoniana* and *Astragalus schimperi* may be greater than the vertical, up to 20 or 30 cm. Most annual species flower by the middle to end of March, dry up and set seed.

In a transect with such marked distinctions between one season and another, seasonal changes in the floristic composition of the vegetation may be significant, and the longevity, or the annual, biennial or perennial habit has to be considered. In the arid zone of Rajasthan, it is possible to divide plants into four classes according to their longevity.[197]

PERENNIALS

1. *Cenchrus ciliaris*. 2. *C. setigerus*. 3. *Dactyloctenium sindicum*. 4. *Eleusine compressa*. 5. *Cyperus rotundus*. 6. *Boerhavia diffusa*. 7. *Citrullus colocynthis*. 8. *Convolvulus pluricaulis*. 9. *Tephrosia purpurea*.

ANNUALS, BIENNIALS OR PERENNIALS

1. *Aristida adscensionis*. 2. *Cenchrus biflorus*. 3. *Digitaria adscendens*. 4. *Cyperus arenarius*. 5. *Tribulus terrestris*.

ANNUALS

1. *Eragrostis ciliaris*. 2. *Perotis indica*. 3. *Indigofera cordifolia*. 4. *I. linifolia*. 5. *Heliotropium strigosum*. 6. *Justicia simplex*.

EPHEMERALS

1. *Eragrostis pilosa*. 2. *Tragus biflorus*. 3. *Urochloa panicoides*. 4. *Corchorus aestuans*. 5. *Gisekia pharmacioides*. 6. *Heliotropium paniculatum*.

The perennial nature of the first nine species is recognized by all botanists, but the biennial nature of the other five species listed is a matter of controversy. These species behave both as perennials and as biennials and rarely as annuals depending upon the favourableness of their environment. This longevity may be an adaptation to a desert climate and arid soils, the light texture of which allows their roots to go deeper for better establishment and to withstand the unfavourable periods. There is little doubt regarding the annual nature of the six species listed, but the ephemeral nature of the other six species varies with their environment. Competition plays an important part; a plant facing severe competition for its requirements in a dense vegetation behaves like an ephemeral by completing its life cycle within a few days, while individuals of the same species prolong their life cycle for a longer period when growing in areas devoid of competition during the favourable season.

The longevity of the four groups of species depends upon their water requirements. Roots also play an important role, because the nature of their establishment in the soil profile is responsible for the availability of water and nutrients. It may be broadly concluded that the deeper the roots of plants, the longer is the period of their life cycle.

As regards the sociological variations in the floristic composition of the monsoonal vegetation of the alluvial plains of the arid zone in Rajasthan, five species occur in summer: the perennials, *Cenchrus setigerus* and *Eleusine compressa*, are dominant and remain in a living condition up to April. Three weeks after the first showers, the number of species increases to 24 and a fortnight later to 26. During the monsoon period from 30th July to 22nd September, all perennials except *Cenchrus setigerus* increase in the number of colonies but decrease in the number of individuals. Among the annuals, *Cenchrus biflorus* and *Digitaria adscendens*, *Aristida adscensionis* and *Eragrostis ciliaris* behave like perennials. Forbs rapidly increase in number, but perennial weeds germinate more slowly. Ephemerals like *Urochloa panicoides* and *Tragus biflorus* do not stand competition with other species, whereas weeds compete successfully with the others. Changes in the horizontal direction are accompanied by relative increase in the vertical direction. *Eleusine compressa* proves to be the best growing among all the grasses of the alluvial plains. The highest percentage cover and density of annuals are found during the middle of the monsoon, while there may not be any increase in the density of perennials after the middle of the monsoon. Mortality (percentage) in annuals is due to their short life period and poor establishment and lack of competitive power. *Aristida adscensionis*, *Cenchrus biflorus*, *Digitaria adscendens*, *Cyperus arenarius* and *Tribulus terrestris* behave sometimes as annuals, or as biennials or perennials depending on the prevailing conditions.[197]

Savannahs

Anyone concerned with monsoonal grasslands must necessarily take account of the current discussions of the correct nomenclature and the ecological position or status of savannahs in the overall vegetational picture. Many of the associations of our various types of grass covers in woodland, forest or similar communities may be grouped under the somewhat controversial name of 'savannah'.

Foresters and grassland ecologists tend to be perhaps over-influenced by biotic and anthropogenic factors, and always ask themselves first whether a new vegetation type which they have found is not a seral stage in the succession. They tend to think first in terms of dynamic ecology and never take a vegetation at its face value. Hence we find discussion of questions such as the status of the Nilgiri grasslands reviewed in Chapter VI comes in this category. In Ceylon, there is no special reason, according to C. H. Holmes,[91] at any rate of a macroclimatic nature, why the climax vegetation should anywhere be other than closed forest of evergreen or mixed evergreen type, except where it is edaphically or biologically conditioned. Yet in the hills and the plains, in the wet and the dry zone, in fact almost everywhere, are to be found areas, from less than an hectare to seemingly endless tracts, of vegetation of other than closed-forest type, composed of savannah, grass and farmlands.

Savannah in Papua and New Guinea has an open tree storey and a ground cover of grasses, and is structurally similar to the woodlands of northern Australia.[90] It is not possible to distinguish savannah from grassland by stating a given number of trees per unit area. 'In a savannah, the trees dominate in the impression which the landscape makes on the observer; in a grassland the grasses are the most striking feature and the trees do not obstruct the view over the landscape. However, there will always be cases in which a decision is difficult, and here floristic affinities may help in classification.' Savannah in Papua and New Guinea occurs in a range of environments, but mostly in the coastal hill and foothill zones. Both the grasses and the trees have a high resistance to fire. Straight boundaries between savannah and semi-deciduous thicket, which do not coincide with changes in soil or topography, suggest that thicket is replaced by savannah through human interference. On the other hand, boundaries may sometimes be correlated with such changes; on crests of ridges in the foothills, which are otherwise forested, it appears that the occurrence of adjacent forest and savannah is a natural phenomenon.

M. M. Cole has approached the problem as a 'biogeographer',

in her Presidential Address to the 1963 meeting of the South African Geographical Society.[50] In place of the terminology adopted in the preparation of the AETFAT map[4] of the vegetation of Africa south of the Tropic of Cancer, and in place of that used for Australia by R. J. Williams[258] and J. G. Wood[259]—see Table 5—she proposes the new terminology indicated in Table 4. Further reference to existing methods of classifying the vegetation of Australia and to Cole's proposed revision of these methods is made in Table 32. There is greater comparability between the vegetation of Africa and Australia than of either with South America. In both Africa and Australia, Cole's five savannah sub-formations are represented and are distinct from one another floristically as well as structurally and physiognomically. South America, on the other hand, has two distinct suites of savannah vegetation, the one characterized by grasses and the other by thorny succulents.[50]

TABLE 4

Savannah terminology proposed by M. M. Cole[50]

Savannah woodland

Deciduous and semi-deciduous woodland of tall trees (more than 8 m. high) and tall mesophytic grasses (more than 80 cm. high); the spacing of the trees more than the diameter of canopy.

Savannah parkland

Tall mesophytic grassland (grasses more than 80 cm. high) with scattered deciduous trees (less than 8 m. high).

Savannah grassland

Tall tropical grassland without trees or shrubs.

Low tree and shrub savannah

Communities of widely spaced low growing perennial grasses (less than 80 cm. high) with abundant annuals and studded with widely spaced low growing trees and shrubs often less than 2 m. high.

Thicket and scrub

Communities of trees and shrubs without stratification.

The terminology of plant formations has been the subject of international meetings at Yangambi in the Congo in 1955, and at Abidjan in the Ivory Coast in 1959. UNESCO held a Symposium on Tropical Savannahs in Venezuela in 1966, but with emphasis on problems such as the ecological regression of tropi-

A classification for savannah vegetation proposed by M. M. Cole

Proposed Classification	AFRICA		AUSTRALIA		
	Vegetation Association	Authority R. W. J. Keay *et al.*[4]	Vegetation Association	Authority: R. J. Williams[258]	Authority: J. G. Wood and R. J. Williams[260]
Savannah woodland	*Brachystegia-Isoberlinia-Julbernardia* woodlands	Woodlands, savannahs (and steppes)	*Eucalyptus* woodlands	Woodland	Savannah woodland
	Cryptosepalum low forest and woodland	Woodlands, savannahs (and steppes)			
	Baikiaea plurijuga woodlands	Dry deciduous forest and savannah	*Callitris glauca* woodland	Low-layered woodland	Savannah woodland
	Colophospermum mopane woodlands	Woodlands, savannahs (and steppes)	*Acacia harpophylla* woodlands	Low-layered forest	Savannah woodland or savannah
Savannah parkland	*Acacia-Terminalia-Piliostigma-Combretum* grasslands	Woodlands, savannahs (and steppes)	*Acacia-Bauhinia-Terminalia* grasslands	Tree and low tree savannah or included in tussock grassland	Savannah
Savannah grasslands	*Hyparrhenia-Themeda-Setaria-Echinochloa* grasslands	Swamps	*Astrebla-Iseilema-Dichanthium* grasslands	Tussock grasslands	Savannah
	Trichopterys grasslands	Grass steppe			
Low tree and shrub savannah	*Chrysopogon-Aristida-Cenchrus* grasslands with *Acacia* and *Commiphora* spp.	Wooded steppe with abundant *Acacia* and *Commiphora*	*Triodia* associations with *Acacia* and *Eucalyptus* spp.	Tree and low tree savannah	Desert steppe
				Shrub savannah Hummock grassland	
Thicket and scrub	*Acacia-Commiphora* thickets		*Acacia aneura* scrub	Low-layered woodland	Mulga scrub
			Acacia shirleyi scrub *Acacia cambagei* scrub	Low-layered woodland	

cal forests to the savannah stage, laterization, etc., rather than on correct and internationally accepted nomenclature. Those working in monsoonal ecoclimates can but agree with M. M. Cole on the need to achieve such a classification of the vegetation of the tropics made by plant ecologists and geographers, with structure, physiognomy and floristic composition being given the correct degrees of emphasis.

Altitude

Interesting questions relating to the occurrence of grass species at high altitudes in our transect are raised in the chapters on India, Taiwan, Malesia (Figs. 47–50) and Papua and New Guinea. Are these relict areas of temperate grass genera, and sometimes also species, in habitats that are widely separated geographically, an indication of greater degree of continuity in some earlier climatic age? This is a large subject that calls for a special review to itself. In West Africa there is considerable discussion regarding the Cameroon Mountain flora, following A. Chevalier[36] and R. Good.[76] R. E. Moreau[152] has suggested that this mountain is of Pleistocene origin. J. K. Morton[153] in his study of the temperate flora of West African mountains proposed three groups for highland Cameroons—East African Montane, Asiatic Temperate and Boreal. F. N. Hepper[89] states that the flora of the Cameroon highlands must be regarded as an impoverished flora of highland East Africa. We await the report of the Symposium held in Mexico City on 1st to 3rd August 1966 on the geo-ecology of the mountainous regions of the tropical Americas.

1. Rainfall gradient on sandy soils. Savannah with tall grass and scattered trees. (*Photo Ch. Rossetti*)

2. Plant formation following destruction of former savannah for cultivation of groundnuts. (*Photo Ch. Rossetti*)

[For a detailed description of these plates see Appendix 2]

3. Predominantly annual cover, due to edaphic factors. Brown soil is loamy sand. (*Photo Ch. Rossetti*)

4. Characteristic physiognomic variant of vegetation in Plate 3, on more compact sand. (*Photo Ch. Rossetti*)

[For a detailed description of these plates see Appendix 2]

5. Short-duration annual sward with small trees and shrubs.
(*Photo Ch. Rossetti*)

6. Effect of biotic factors on preceding community.
(*Photo Ch. Rossetti*)

[For a detailed description of these plates see Appendix 2]

7. Steppe grasslands showing transition between annual grasslands to south and continuous steppe cover to north. (*Photo Ch. Rossetti*)

8. Steppe of *Panicum turgidum* with annuals, and some trees and shrubs at their northern limit. (*Photo Ch. Rossetti*)

[For a detailed description of these plates see Appendix 2]

Chapter IV

AFRICA

WEST AFRICA—SAHARA/SAHEL

Climatic gradient

It is well-known that a major feature of West Africa is a north/south gradient of decreasing aridity which is apparent across the African continent between 10° and 20° North latitude, from the Atlantic to the Red Sea coasts, or perhaps more correctly to the Nile, with certain major orographical exceptions. A. Aubreville[5] has proposed a subdivision into bioclimates and recognizes the Sahara, Sahelo-Saharan and Sahelo-Sudan climatic zones. Charles Rossetti[191] has followed the same subdivision in his meticulous study (unfortunately available only in mimeographed form and hence quoted rather fully here) of the natural environment of West Africa in relation to the breeding grounds of the desert locust (*Schistocerca gregaria*) (Fig. 11). The subdivisions do not, however, correspond to those of the plant geographers, who employ much the same terms but with rather different connotation.

Geobotanical gradient

The classic geobotanical subdivisions are without exception referred to an interpretation of the climatic gradient, reduced to terms of rainfall. All authors who have adopted this approach are agreed about the existence of three latitudinal zones: Sahara, Sahel and Sudan (A. Chevalier,[37] B. Zolotarevsky and M. Murat,[264] Th. Monod,[149, 150] M. Murat,[155] R. Capot-Rey,[28] A. Aubreville[5]). There is, however, considerable divergence of opinion regarding the precise lines of demarcation between the

TAB[…]

Ecoclimatic gradient in western a[…]

Zones	Approximate rainfall range, mm	Dry season, months	Vegetation (Keay, [4])	Forest formations	Chief tree species	Grass cover types (Rattray, 18[…]
Saharan			Desert	—	—	*Aristida* (12); desert conditions
Sub-Saharan	100–250		Sub-desert steppe; tropical types	—	—	*Aristida* (13, associated wi[…] steppe
Sahelian	250–600	9	Wooded steppe with abundant *Acacia* and *Commiphora*	Shrub and thorny savannah, Sahel types	*Acacia* *Commiphora*	*Cenchrus* (5, 6, 7) ass. with savanna[…]
Sudanian	600–1,250	6–8	Woodlands, savannahs, undifferentiated, relatively dry types	Open woodland (tree savannah), Sudanese types	*Anogeissus* *Sclerocarya* *Balanites* *Prosopis* *Butyrospermum* *Adansonia* *Bombax*	*Andropogon* (3, 5, 6) ass. with savannah
Guinean	above 1,250	3–6	Woodlands; *a.* Northern areas with abundant *Isoberlinia doka* and *I. dalzielii* *b.* Southern areas, undifferentiated, relatively moist types	Seasonal (deciduous) forest, Guinean types	*Isoberlinia* *Berlinia* *Uapaca* *Lophira* *Brachystegia*	*Hyparrhenia* (22, 31, 33, 3[…] 35, 36, 37) as[…] with savanna[…]
Guinea Equatorial	above 1,800	short	Moist forest at low and medium altitudes	Closed rain forest	*Khaya* *Entandrophragma* *Lovoa* *Piptadenia* *Lophira* *Mytragina* *Sacrocephalus* *Aucoumea* *Triplochiton* *Tarretia* *Chlorophora* *Terminalia*	*Pennisetum* (5, 6, 7) ass. with woodland

stern equatorial regions of Africa[253]

ıin types of land-use	Main crops	Livestock	Countries	Zone
madic grazing	—	Some camels	Mauritania, Mali, Niger, Chad, Sudan	**Saharan**
madic grazing	—	Camels, sheep and some cattle	Mauritania, Mali, Niger, Chad, Sudan	**Sub-Saharan**
ni-nomadic grazing h beginnings of semi- d cultivation. Minor e products and tection against iccation	Sorghum and millet, irr. rice	Cattle and sheep	Senegal, Mauritania, Mali, Upper Volta, Niger, Chad, Sudan	**Sahelian**
ni-nomadic grazing l arable cultivation h fallows generally of dium to long dura- n. Woodland utiliza- n for minor products	Sorghum, millet, groundnuts, yams, maize, irr. rice	Cattle with some sheep and goats	Senegal, Mali, Upper Volta, Niger, Northern Nigeria, Chad, Sudan	**Sudanian**
ıble cultivation with ows generally of dium to long duration. rest utilization for al consumption. rest plantations	Sorghum, millet, groundnuts, cassava, yams, maize, irr. rice	Cattle depending upon absence of tsetse, goats and some sheep	Senegal, Gambia, Portuguese Guinea, Guinea, Mali, Ivory Coast, Upper Volta, Ghana, Togo, Dahomey, Nigeria, Cameroon, Chad, Central African Rep., Sudan, Congo (Brazzaville), Congo (Kinshasa)	**Guinean**
ıble cultivation with ows generally of ıtively long duration. manent tree crops for ort. Forest utilization export and local ısumption (natural nds and plantations)	Upland rice, plantains and bananas, yams, taro, maize, oil palm, cacao, rubber, robusta coffee	Goats, with cattle in forest-savannah mosaic on coast	Sierra Leone, Liberia, Ivory Coast, Ghana, Nigeria, Cameroon, Rio Muni, Gabon, Congo (Brazzaville), Cabinda, Congo (Kinshasa)	**Guinea Equatorial**

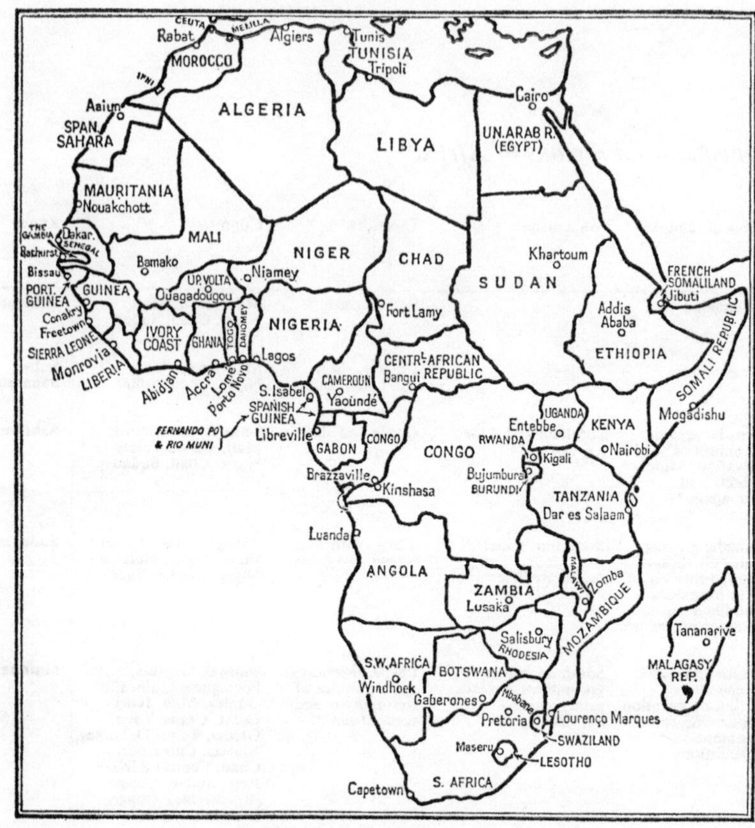

FIGURE 18
Political map of Africa

different zones, and regarding the desirability of postulating a fourth zone between the Sahara and the Sahel.

A schematic presentation of this classic zonation of West Africa is presented in Table 6,[253] indicating a certain type of climatic and vegetation zonation carried through to its utilization aspects in terms of grazing patterns, types of animal husbandry and cropping.

Seasonal climate

The seasonal climatic regime of the monsoonal zone of West

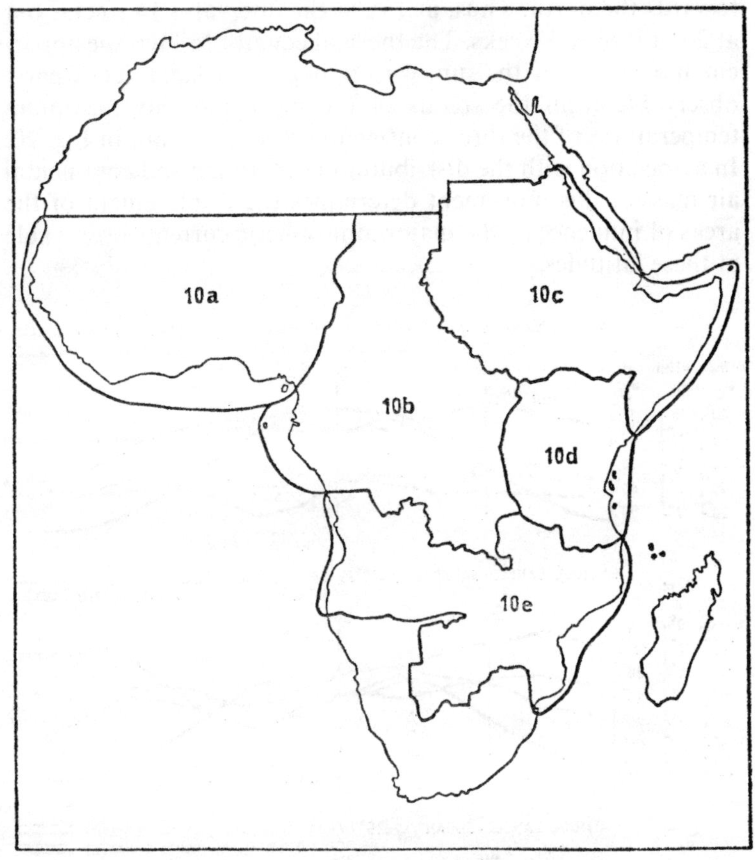

FIGURE 19

The geographical subdivisions of tropical Africa in use at the Royal Botanic Gardens, Kew

Key

10a. West Tropical Africa
10b. Cameroons and Congo
10c. North-east Tropical Africa
10d. East Tropical Africa
10e. South Tropical Africa

Africa[191] in which our transect begins consists of a short rainy season and a long dry season. The seasonal character is correlated with the apparent movement of the sun towards the zenith, through which it passes twice a year at particular dates in each latitude. The interval between the two passages lessens towards the north. Thus, at 15° N, the interval is 14 weeks, but at 20° it is only 5 weeks. The thermal equator follows the apparent movement of the sun, with only a slight lag, a fact clearly observable from the trends of the mean monthly maximum temperatures of the three continental regimes set out in Fig. 20. In association with the distribution of maritime and continental air masses, this movement determines the displacement of the areas of influence of the major atmospheric currents observable at these latitudes.

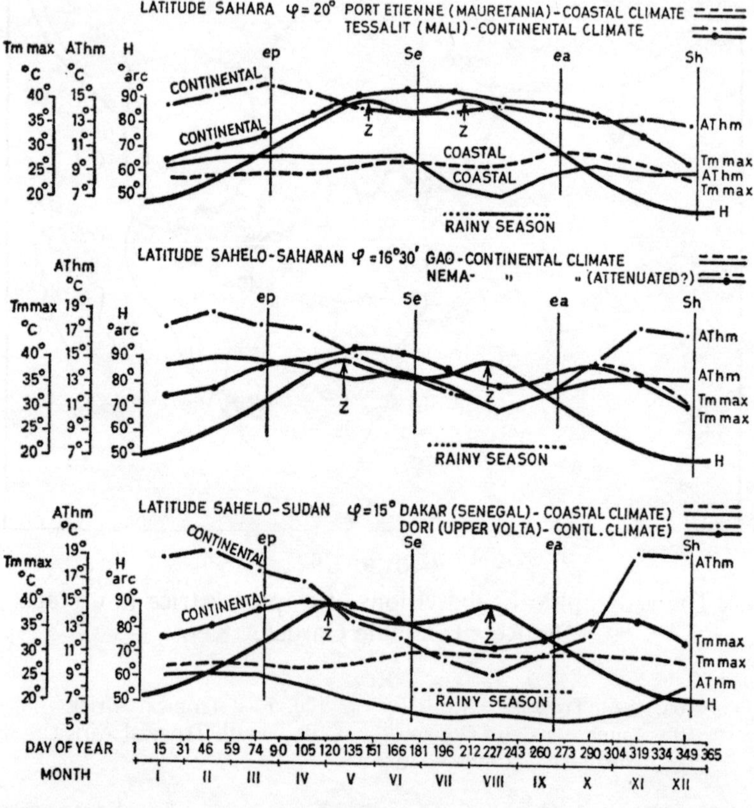

FIGURE 20

Altitude of sun at transit through local meridian, and annual temperature pattern at latitudes 15°, 16°30′ and 20° N, after Ch. Rossetti[191]

In winter the Sahara is a high-pressure area where the northerly continental trade wind (Sahara current) has its origin. This is a hot, dry wind capable of appearing east of the Greenwich meridian as far south as about 5° N, while west of that meridian it is only rarely felt south of 15° N. In summer the Sahara, greatly heated owing to the position of the sun, becomes a low-pressure area and draws in humid maritime equatorial air. The zone between these two air masses, the inter-tropical convergence zone (ITCZ), accordingly shifts from south to north, a phenomenon during which the mechanisms giving rise to rainfall are triggered off. While the ITCZ is for the most part displaced towards the north, the maritime equatorial air masses push to a higher latitude the third major air current blowing throughout the year from east to west, viz. the hot and dry continental equatorial air current (or 'harmattan'). The contact between the warm, humid coastal air current, and the continental current, now cooled through having been pushed upwards, causes squall lines to form, giving rise to storms which may be either dry (in which case they do no more than raise clouds of dust) or bring a heavy shower. From south to north the phenomenon becomes less frequent, particularly as regards the accompanying showers.

The harmattan, sweeping down suddenly from high altitudes, brings cool squalls in its train and causes temperatures to drop with corresponding suddenness. Thus, during the rainy season in the Sudan and Sahelo-Sudan latitudes, the mean monthly maxima are some 10° lower than the maximum figure reached shortly after the first passage of the sun through the zenith during the rainy season (see Fig. 20), a phenomenon due to the erratic character of the rains in these latitudes. In the coastal regime, the seasonal swing of the prevailing wind occurs between the Atlantic trade winds and the monsoon with a gradual seasonal attenuation of the monsoon as one proceeds from south to north.

While the continental regimes are characterized at all latitudes by a single rainy season occurring regularly at the period when the sun passes for the second time through the zenith, to the west of 10° W the Saharan latitudes may receive some rain in winter/spring. The effect of this regime would decrease from north to south and from west to east. However, this question cannot be treated in greater detail, important though it may be to distinguish the ecological conditions obtaining in the normally dry season in what one might call the 'Atlantic' Sahara from conditions characterizing the continental Sahara.

Zonation of plant cover

In a survey in east Mali in 1959, Rossetti[191] observed the plant cover zonation in the meridional zones. Starting with the work done at Yangambi by the Commission for Technical Co-operation in Africa (1956), on which the AETFAT[4] vegetation map of Africa was based, Rossetti found it possible to arrive at certain criteria for classification which are set out below.

Throughout the areas surveyed, the plant associations consist of mixtures of varying proportions of herbaceous and woody plant strata. In the zonation of the woody stratum, there are, from south to north, a more or less steady decrease in the density and bulk of individual plants, and numerous intrusions in the distribution of floristic elements. To adopt a term used in a different context by Monod[150] the zonation in question forms a *continuum*. A succession of this kind scarcely lends itself to the establishment of units or the definition of boundaries. The herbaceous strata are for the most part made up of grasses. In contrast to the woody strata, the grass covers are differentiated by zonal characteristics which are seldom found in more than one zone. On the ground, the transition from one type of cover to another is often marked by a clearly discernible line of separation or is at least confined within a narrow compass. The succession of such lines may be correlated with the climate gradient over one and the same substratum (in this case, sand).

The species making up the grass cover may be classified into three types according to their biological cycle, the characters of the leaves and their position on the plant, their habit, and the height of the flowering culms, thus:

(i) prairie made up entirely of annual grasses mostly under 80 cm. in height, uniformly distributed, with aerial portions often forming a continuous cover. The habit is generally erect, occasionally prostrate; the leaves narrow;

(ii) steppe characterized by tussocky perennial grasses having an apparently uniform distribution. The grasses have narrow

9. Further north, replacement of *Panicum turgidum* by *Aristida pungens*. (*Photo Ch. Rossetti*)

10. Edaphic communities in the region with less than 150 mm. rainfall. (*Photo Ch. Rossetti*)

[For a detailed description of these plates see Appendix 2]

11–13. Edaphic co
munities in the re
with less than 150
mm. rainfall.
(*Photo Ch. Rosset*

[For a detailed description of these plates see Appendix 2]

leaves, mainly basal. The habit is semi-erect, semi-prostrate. The culms are less than 80 cm. tall;

(iii) savannah consisting of tussocky perennial grasses having an apparently uniform distribution. At flowering time the species are of erect habit and the culms usually over 1·5–2 m. in height. The leaves are thick, not rolled, basal and cauline. The aerial parts form a virtually continuous cover.

The following species are typical of the three types of grassland:

EPHEMERAL PRAIRIE	STEPPE	SAVANNAH
Aristida funiculata	*Aristida pungens*	*Andropogon gayanus*
Aristida mutabilis	*Aristida pallida*	*Cymbopogon giganteus*
Cenchrus biflorus	*Aristida acutiflora*	*Hyparrhenia dissoluta*
Schoenefeldia gracilis etc.	*Aristida papposa*	
	Cymbopogon schoenanthus	
	Eremopogon foveolatus	
	Panicum turgidum etc.	

While there is a regular interpenetration of the prairie grassland with the steppe and savannah, a mixture of steppe and savannah species was observed only once.

There is also a fourth herbaceous type found at sites which are highly specialized from the soil standpoint and are not very extensive, namely grassland areas made up of the perennial hygrophilous grasses usually exhibiting a prostrate or semi-erect habit (*Cynodon dactylon*, *Sporobolus helvolus* and others).

The woody strata comprise individual plants whose aerial parts exhibit a high degree of lignification. They are subdivided according to the height of the component plants and the branching of their stems. Independently of the height of the individual plants, the term *bushy* is applied to these woody layers made up of multicauline plants with the stem branching from the base up to a maximum height of about 50 cm. Plants having a taller trunk are shrubs (up to 5 m.) or trees (when the total height is over 5 m.). The term *chamaeophyte* as defined by Raunkiaer (shoots up to a maximum height of 25 cm. above ground) is retained because this stratum, which is very loose, is characteristic of many Saharan associations (*Moltkia ciliata*, *Indigofera argentea*, *Crotalaria saharae*, etc.).

Zonal and local characteristics of vegetation

Within the region studied were noted three regions in particular:

(*a*) Along the northern boundary of the Sahelo-Saharan region, the most striking type of terrain is the savannah formations which are suddenly encountered as one travels from the north, at first on sandy soils, and generally in patches from a few hundred to a few thousand hectares in extent, with an annual rainfall around 350–450 mm. The characteristic stratum consists of a continuous cover of tall grasses (*Andropogon gayanus* or *Hyparrhenia dissoluta*) which may grow 2–2·5 m. high at flowering, when the aerial portions form a virtually continuous cover. This situation is associated with a low herbaceous layer of annual and perennial species of *Aristida, Cenchrus, Eragrostis* and *Cyperus*. This savannah vegetation is herbaceous, although in Mauritania Rossetti noted *Sclerocarya birrea* and *Combretum glutinosum*, and in eastern Mali *Guiera senegalensis* and a few *Combretum glutinosum*. Murat[154] has noted some steppe elements due to slight ecological variation of a site, for example, *Aristida stipoides, Cenchrus biflorus, Eragrostis tremula* and *Panicum turgidum*. Rossetti queries whether the great extent of ephemeral grasslands in the Sahelo-Saharan region (below) reflects a general degradation in the savannah zone. 'Are the savannah types that are characteristic of this northern edge of the Sudan region vestiges of a type of savannah that may once have covered a wider area? Or do they merely represent the ramifications of a Sudanese physiognomy on substrata with an especially good water balance?'

(*b*) North of the Sahelo-Sudan border, which is characterized by patches of savannah on flattened dune relief, the tall grass stratum is no longer in extensive and widespread stands and occurs only in hollows with a fair water supply. In one area, a savannah type grass, *Cymbopogon giganteus*, is found occasionally but never in continuous stands. In this zone Rossetti recognizes the 'ephemeral prairies', which cover one degree of latitude in extent, and continue as far as those areas where the steppe appears further to the north. In the south of this belt there are no perennial species, but to the north there are patches of perennial grasses with the annual stratum sufficient to justify the term 'steppic prairies'.

In the ephemeral prairies, *Cenchrus biflorus* is the most abundant species, along with *Aristida mutabilis, A. stipoides, Schoenefeldia gracilis, Dactyloctenium aegyptium, Tragus racemosus* and *Latipes senegalensis.* In the woody strata there is a tendency for the species to associate in a south–north manner, according to the following sequence: south of the 300 mm. isohyet, mainly *Balanites aegyptiaca* (shrubs), *Commiphora africana, Maytenus senegalensis, Combretum glutinosum, C. aculeatum*; north of the 300 mm. isohyet, mainly *Balanites aegyptiaca* (shrubs and bushes), *Acacia raddiana, A. senegal, Leptadenia pyrotechnica, Acacia laeta.* However, this is no more than a tendency. When surveyed in the field, distributions are much more confused and their boundary lines correspondingly vague. Also the sequence may obviously be disturbed by such factors as relief, drainage, soil conditions and sand formations.

The closed drainage, especially in east Mali (Gourma), gives rise to marked contrasts, first of all in the mode of distribution of the woody stratum. This becomes more dense and gives rise to wooded patches in lower-lying parts or around pools (*Acacia nilotica, Anogeissus schimperi, Hyphaene thebaica, Mitragyna inermis* or thickets of *Acacia flava* in the north; and *Euphorbia balsamifera, Boscia senegalensis, Zizyphus mauritiana, Acacia seyal, Combretum micranthum, Piliostigma reticulata, Bauhinia rufescens* and *Dichrostachys glomerata* in the south).

According to their position in the drainage system (centre or edges of pools, or main course of drainage) the following, largely hygrophilous species are observed: *Echinochloa stagnina, Panicum laetum, Fimbristylis exilis, Andropogon gayanus* and others.

The seasonal dynamics of the ephemeral prairies in this Sahelo-Saharan region are closely linked with the rainfall regime; the period of rapid active growth is usually ended before the end of the rainy season. In the 'brousse tigée' of the Dahr de Nema, a longer active-growth period was noted, with *Andropogon gayanus* still partially flowering in late September and early October when most deciduous bush species (*Commiphora* and *Grewia*) had already shed their leaves and the adjacent prairies had dried up completely.

It was therefore surprising to learn of the extent of these

prairies in the Sahelo-Saharan region, and it is reasonable to enquire as to the explanation of this distribution over more than one degree of latitude (approximately between 400 mm. and 300 mm. isohyets). Is it that perennial species are absent? Is their absence due to over-grazing? Here again, before one can answer, one must know what a climax grass cover, or in simpler terms a natural grass cover, protected from fire and grazing, would be like. In the survey area, the entire absence of declared reserves made it impossible to shed any light on this problem. Another question is whether the individual perennial grass species occasionally encountered, though few and far between, in the zone under discussion are to be regarded as vestiges. Among such species in south-east Mauritania may be mentioned: *Andropogon gayanus*, *Aristida pallida* (or *longiflora*?), *Cymbopogon* (*giganteus*), *Schmidtia pappophoroides*, *Aristida papposa*. No perennial grass was recorded in the equivalent zone in eastern Mali. A common instance of our ignorance of ecology, this question is likely to remain open so long as a narrowly floristic or phytogeographic bias continues in research on vegetation.[191]

(*c*) There are two Saharas, the Sahara of sand and the Sahara of pebbles, with obvious distinction in the vegetation. In the sand areas one finds when moving from the south into the northern fringe of the Sahelo-Saharan region that there are in the ephemeral grasslands, at first as isolated plants and subsequently in patches perennial species such as *Aristida pallida* and *Panicum turgidum* forming an open cover with more or less bare stretches of sand between the tussocks of these species. Tracts of this 'steppic stratum' appear first on the higher parts of the sand relief—on sandy plains it may be found distributed diffusely in the ephemeral prairie. Further north the steppe tracts become more extensive and continuous, with patches of annual grasses becoming rare. The location of these tracts and the borders between the physiognomic types are naturally dependent upon a number of local site conditions. For example, in one valley, the dune-lined sides are colonized by a bushy steppe of *Panicum turgidum*, *Aristida pungens* and *Calligonum comosum*, the bottom of the valley by an ephemeral prairie of *Aristida mutabilis* and *Cenchrus biflorus* but also including *Acacia raddiana* and *Balanites aegyptiaca*.

At the southern limit of distribution, these steppes are in all cases of the shrubby or bushy type. There one finds *Balanites*, *Maerua* and *Commiphora* (though this has little northward extension), occasionally *Euphorbia balsamifera* and *Leptadenia*. The most frequently found woody species is *Acacia raddiana*.

As one goes north the gradual disappearance of a number of herbaceous species (*Aristida pallida*, *A. stipoides*, *Cenchrus biflorus*, *Sesamum alatum*, *Aristida mutabilis* etc.) present in the Sahelo-Saharan region is clearly noticeable, and the following begin to appear (or to increase in frequency): *Aristida acutiflora*, *Requiena obcordata*, *Euphorbia scordifolia*, *Indigofera argentea* and, in south-east Mauritania only, *Neurada procumbens* and *Heliotropium* sp. The prostrate chamaeophytes, *Crotalaria saharae*, *Chascanum marrubiifolium*, *Moltkia ciliata* (true chamaeophyte or therophyte?) and *Indigofera argentea*, which are absent further south, seem to be characteristic of the Saharan region. *Cornulaca monacantha*, generally regarded as a species of that region, was not observed during any of the trips in the heavily sanded region in south-east Mauritania. A few specimens were noted in southern Tanezrouft.

Near 17° 40′ N, between Aioun and Tichit (Aouker, south-east Mauritania) and 18° N, between Tombouctou and Bou Djebeha (Asouad, eastern Mali), i.e. in the 150–100 mm. annual rainfall range, is the southern limit of a steppe grass species, *Aristida pungens*, which occupies an increasingly important place northwards, finally replacing *Panicum turgidum* totally. As in the case of *Panicum turgidum* further south, *Aristida pungens* is associated where the rainfall is lower than 150 mm. with the presence of shifting sands, exhibiting a less pronounced relief, however, than at the southern limit of the *Panicum*. In its own area of distribution, *Aristida pungens* again gives way to *Panicum* at places where the sand is shallow (rocky outcrops), or even where it is deep but very compact or structureless (clay fraction over 4 per cent).

Observations (unpublished) made in the Mauritanian Adrar on the distribution of these two species were conclusive. Dunes of wind-formed deposits that pile up against the escarpments are in all cases occupied by *Aristida pungens*, while open tracts made up of sands difficult to classify, often containing fine gravel, have stands of *Panicum turgidum*.[191]

In these vast areas, man's influence on the composition and density of the vegetation is not very noticeable except in the immediate vicinity of wells, where the steppe is entirely eliminated over a fairly large area. On the nebka (mounds of sand) that form around wells as a result of wind action, bushy patches of *Calotropis procera*, *Leptadenia pyrotechnica* and *Acacia raddiana* appear owing to local elimination of perennial herbaceous plants. One occasionally comes across open herbaceous cover, for the existence of which no local ecological reason is apparent. In some cases these variations are accompanied by mutilations of the shrub layer in the form of lopping and topping done by the nomadic drovers for the purposes of feeding their animals, so that the influence of the biotic factor has to be recognized. On sand the most pronounced phenomenon of degradation in this steppe region is, however, unquestionably the replacement of steppe formations by physiognomies of steppic or ephemeral prairies, and this encourages the belief that the latter formation extends both north into the steppe region and south into the savannah as a consequence of the elimination of herbaceous perennials. Here again, the task of proving such degradation calls for time-consuming surveys based on exact measurements, which cannot be attempted in the absence of reserved areas. The findings of surveys (unpublished) conducted in the reserves of the Adrar district (north-west Mauritania), the latter exhibiting some ecological conditions resembling those in the sandy areas discussed here, apparently do not reveal any appreciable increase in the plant mass in the steppe at the end of 6 years under reserve.

Areas with relief within the Saharan region naturally make it impossible to trace the exact impact of climatic zonation on the pattern of ground cover. These areas may have a marked influence upon the rainfall distribution and give rise to a northward intrusion of monsoonal phenomena. Species from extra-Saharan climatic zones may become established (R. Maire and Th. Monod;[137] G. Carvalho and H. Gillet;[29] P. Quezel;[179] R. Maire and B. Volkonsky[138]). It may be true in the Saharan region that the seasonal dynamics are linked with the monsoon rains but less closely. Rossetti makes the tentative observation that, when the rainy season is over, some of the plants in the Saharan herbaceous strata continue in a state of vegetative activity (especially towards the north), when the ephemeral prairies of the Sahelo-Saharan region become inflammable straw mats.

Physiognomic subdivisions in West Africa

Rossetti summarizes his conclusions as follows:

Between latitudes 15° and 18° N, a belt which was traversed in parts between the meridian of Greenwich and 10° W, there are from south to north three vegetation zones which may be distinguished by certain characteristics of the predominant grasses in the herbaceous layer: these are savannahs, ephemeral prairies and steppes.

The composition of certain floristic associations seems to indicate that the distribution of a number of non-graminaceous species characterizing the physiognomy of the vegetation is associated with each of the three zones in that their penetration into a neighbouring physiognomic zone is accidental and often due to the local appearance of particular edaphic conditions.

The floristic and physiognomic zonation thus conceived is evident from field observations conducted along the meridian only in the case of sandy substrata of greater depth than the limit of penetration of the roots of gramineous species (i.e. a few metres). In areas with little sand or where sand formations are intermittent or completely absent, in areas with relief, the soil moisture conditions and the water supply due to runoff affect the characteristics of the gradient to a significant degree.

In many cases the vegetation pattern in these regions, which may extend across several zonal belts, is not comparable with the reference gradient on sandy terrain. Nevertheless, within these regions, the characteristics of the zonal gradient may be found in the complex distribution of the vegetation of certain topographic sequences (wadis and wadi banks, dunes and interdune areas, and places with contrasting soil texture).

Confining consideration to soil conditions of sandy regions, the limits of the subdivisions adopted by certain authors are found to correspond approximately to those of the physiognomic units here postulated. In particular, the 200 mm. isohyet coincides fairly well with the southern limit of continuous steppe tracts. The 400 mm. isohyet marks the appearance of savannah physiognomy in the ephemeral prairies.

Provisionally, the following equivalents are proposed:

Zone of continuous steppe tracts	=	Saharan region
Zone comprising, in the north, a mosaic of ephemeral prairies and steppic prairies and, in the south, ephemeral prairies	=	Sahelo-Saharan region
Zone where savannahs appear in ephemeral prairie	=	Sahelo-Sudan borderline

The influence of biotic factors (cultivation, fire, grazing, etc.) is geographically widespread, but in the absence of reference areas where it could be excluded, it is everywhere difficult to prove, particularly since the biotic factors generally operate (especially in the drier parts of the survey area) in naturally complex ecological pockets susceptible to colonization by 'contrasting' associations in the absence of any interference by man. In the moister part of the survey area, the influence of biotic factors on the vegetation is, by contrast, more marked and easier to determine.

By working on the scale at which climatic zonation comes into play, it was noted that the physiognomic type called ephemeral prairie tends to stretch southwards and northwards at the same time; the grasses with a short growth period (*Aristida mutabilis*, *Cenchrus biflorus*) tend to replace the savannah and steppe grasses which have a slower growing period and certainly in the seedling stage do not easily become established. This phenomenon is particularly evident at the two extremes (in the north for the savannahs and in the south for the steppes), where competition operates particularly against the perennial grasses.

It is tempting to say that the entire belt of ephemeral prairies which lies between the steppe and the savannah owes its existence to human intervention. However, in the present state of our knowledge (or rather ignorance) of the climatic physiognomy of the vegetation of this area, such an assertion cannot be either proved or disproved from the evidence of methodical observation. A continuous study of the regeneration of the vegetation in a series of reserved areas suitably distributed along the climate gradient and in ecologically well-defined stations, would undoubtedly yield some strong arguments in support of our views on the matter.[191]

Classification on basis of types of grass cover

The map of the grass cover of Africa compiled for FAO by J. M. Rattray[183] was the result of the synthesis of a great many published works upon the vegetation in general or the grass cover in particular. As far as possible the preliminary draft of the map and the recognition of types of grass cover were checked with specialists familiar with each country or region. Rattray has classified much of the vast area of Sahara and Sahel under various types all characterized by annual and/or perennial species of *Aristida*. The species of this genus are for our transect what the species of *Stipa* are for the Mediterranean and Irano-Turanian types of vegetation, and also probably for the extensive grass covers of the heartland of Asia—taxonomically advanced grasses which have evolved adaptations to extreme and particularly arid environments and which therefore come lowest in the ecological succession, the last relicts or the first pioneers. Rattray expresses the situation in Africa thus:

> Much of the grass cover of Africa has been so severely maltreated that many of the grasses composing it are secondary, and species of *Aristida* are extremely common in this respect. This is particularly the case in the drier parts of Africa, especially along the fringes of the Sahara, where continuous heavy grazing and trampling by the livestock of nomadic tribes has reduced an already sparse vegetative cover to a condition comparable with the desert itself, and a few scattered plants of *Aristida* are often the only grasses which survive.[183]

Below are given particulars of the botanical composition and geographical location of the five types of grass covers in West Africa in which Rattray recognizes various species of *Aristida* as characteristic, not necessarily ecologically dominant.

A.13 Central African Republic, Republic of the Congo, Gabon Republic, Republic of Chad, Republic of the Niger

Aristida mutabilis—Panicum turgidum—Cymbopogon giganteus —Eragrostis tremula—Cenchrus biflorus (all on sandy soils), and *Aristida funiculata—A. adscensionis—Schoenefeldia gracilis—Schizachyrium exile* (all on clay soils)

These grasses are associated with a thorny bush steppe largely of *Acacia seyal*, *A. tortilis* subsp. *raddiana* and *Combretum glutinosum*. The rainfall is 500–750 mm.

A.14 Central African Republic, Republic of the Congo, Gabon Republic, Republic of Chad (Tibesti), Republic of the Niger (Air), Sudanese Republic (Adrar), Algeria (Ahaggar)

Aristida acutiflora—A. adscensionis—A. funiculata—A. meccana —A. mutabilis—A. plumosa (and several other species of *Aristida*)*—Cenchrus biflorus—C. ciliaris—Enneapogon brachystachyus—Eragrostis pilosa—Lasiurus hirsutus—Panicum turgidum—Pennisetum* spp.—*Sporobolus* spp.

These grasses are associated with a scattered tree steppe of *Acacia* spp., *Dichrostachys cinerea*, *Balanites aegyptiaca*, *Zizyphus* sp. and others, including numerous shrubs. This type is found on the Tibesti and Air mountains at altitudes varying from 500 to 3,200 m. and with a rainfall of 110–120 mm. or more in places, during May to September. A similar grass cover is likely to occur on the Adrar and Ahaggar mountains.

A.15 Sudanese Republic, Mauritanian Islamic Republic

Aristida pungens—A. meccana—A. funiculata—A. mutabilis—A. adscensionis—A. acutiflora—A. papposa—A. plumosa—A. foëxiana—Panicum turgidum—Lasiurus hirsutus—Latipes senegalensis —Pennisetum divisum—Cymbopogon schoenanthus—Cenchrus ciliaris—Cenchrus sp.—*Enneapogon brachystachyus*

These grasses are associated with a tree steppe of vegetation consisting mostly of small scattered *Acacia* trees.

A.16 Morocco, Algeria, Tunisia, Libya, Egypt

Aristida plumosa—A. obtusa—A. acutiflora—A. ciliata (a very palatable species)—*A. pungens* (on the dunes)

(In Libya, *Stipa lagascae*, *S. capensis* and *S. barbata* have also been recorded from this zone.)

These grasses are associated with a steppe type of vegetation consisting of small scattered trees of *Acacia* and dwarf shrubs. *Stipa* sp., *Genista* and *Artemisia* are often found in the 'wadis'. *Lygeum spartum* is common on gypseous soils but not on desert sands. It occurs at altitudes from sea level to 1,000 m. with a winter rainfall of less than 200 mm.

A.17 Morocco, Algeria, Tunisia, Libya, Egypt
Aristida spp.—*Panicum turgidum*—*Lasiurus hirsutus*—*Pennisetum divisum*
(The last three species are almost confined to the 'wadis'.)

These grasses are characteristically found in the Sahara, associated at times with scattered *Acacia* species and shrubs such as *Anabasis*, *Haloxylon* and *Cornulaca*. Extensive areas, however, are devoid of vegetation and only an ephemeral cover of grass and herbs appears after rain.

To the south of this vast territory dominated by species of *Aristida* as a probable indication of intense misuse and desiccation, Rattray finds a belt in which the characteristic grass is *Cenchrus biflorus*, the rather useless annual species of *Cenchrus* easily identifiable by its sharp, spiny burrs that reduce the selling price of wool. This species 'characterizes a zone of steppe vegetation which exists mainly under a 20-inch rainfall', fringing the desert right across Africa from Mauritania to the Sudan. 'It appears to be a pioneer of disturbed areas and remains for long periods in old cultivations. As this zone . . . is grazed heavily every year without any rest periods for recovery, it is difficult to assess what the more permanent species should be.'

In West Africa, Rattray recognizes three types based upon *Cenchrus biflorus*. Included under his type CE.5 for the Republic of Chad are extracts from an FAO report by J. N. Cougoulis.[55]

CE.5 Central African Republic, Republic of the Congo, Gabon Republic, Republic of Chad
Cenchrus biflorus—*Eragrostis tremula*—*Andropogon pseudapricus*—*Panicum subalbidum*

The first two species occur on sandy soils and the latter two on clay soils. These grasses are associated with an open savannah of *Acacia* and *Commiphora* that is approaching steppe in density.

In discussing the improvement of the main cattle routes in the east of Chad, Cougoulis[55] notes the composition of the grass covers.

In the Sahelian zone, in the regions of Batha and north Ouaddai, the shrub savannah dominates the grazing lands. Almost everywhere the vegetation is composed of a closed herbaceous stratum and a shrub stratum, varying widely in appearance from one place to another. In general from north to south, the shrub cover increases progressively in density, and there are also considerable changes in density and botanical composition of the herbaceous cover. In the plain are the common genera of the Sahelian region: *Aristida, Eragrostis, Sporobolus, Schoenefeldia, Cenchrus, Brachiaria, Ctenium*, together with legumes such as *Tephrosia, Indigofera*. In depressions, along the wadis and around the marshes, more hygrophilous and palatable grasses such as *Panicum, Echinochloa, Dactyloctenium aegyptium*, etc., appear.

In the Sahelian-Sudanian zone, in the regions of Guéra and Salamat, the savannah progressively becomes a woodland savannah. The first grasses with Sudanian affinities appear, first scattered, then more and more numerous. These include *Chloris, Pennisetum* and *Andropogon gayanus*. On the heavy clay soils are to be found *Rottboellia exaltata, Setaria* and *Alysicarpus*. On the fairly sandy, poor soils of the Naga occur *Aristida, Schoenefeldia* and *Schizachyrium*.

In the Sudanian zone of Chad which begins a little to the north of Melfi, Djebren, and Ain Timan, the savannah is more clearly wooded, and the density of the tree stratum increases progressively towards the south. The woodland becomes dense along the wadis and the bahres. The plants of *Andropogon gayanus* become increasingly numerous. Here are found *Hyparrhenia rufa* and *Pennisetum pedicellatum*, mixed with *Borreria* and *Rottboellia exaltata*, legumes such as *Indigofera* and *Crotalaria*, and *Vetiveria* and *Oryza barthii* on the sides of marshes. *Ctenium newtonii* and *Loudetia hordeiformis* are predominant on sandy soils, *Beckeropsis* and *Panicum* on wetter areas.

Cougoulis concludes that the value of the pastures on the grazing lands along the routes of animal movement, and above all at watering points, is good. Unfortunately in spite of this, there is great shortage of forage on certain grazing lands because of the intense brush fires which occur in the dry season.

CE.6 Nigeria

Cenchrus biflorus—Eragrostis tremula—Pennisetum pedicellatum —Diectomis fastigiata—Ctenium elegans—Aristida longiflora— Andropogon gayanus

Extensive areas of swamp land occur, characterized by *Vetiveria nigritana, Oryza barthii* and *Echinochloa stagnina*. Much of this area appears originally to have been woodland, but interference by man has opened it out to a savannah type dominated by species of *Combretum*, *Terminalia* and *Acacia*, and even the grass cover is now largely secondary. It includes part of the 'Sudan Zone' and the 'Sahel Zone' of Nigeria. The grasses are usually mostly annuals but 6 years' protection from grazing and cultivation results in its reversion to *Andropogon gayanus*. Rainfall is less than 750 mm.

CE.7 Sudanese Republic

Cenchrus biflorus—Chloris prieurii—Ctenium elegans—Eragrostis tremula—E. pilosa—Aristida adscensionis—A. mutabilis—A. longiflora—A. stipoides—Perotis patens—Latipes senegalensis— Tragus racemosus—Brachiaria hagerupii—B. deflexa—Trichoneura mollis—Pennisetum pedicellatum—Dactyloctenium aegyptium

(Other countries likely to have a similar grass cover, but for which precise information is lacking: Voltaic Republic, Mauritanian Islamic Republic, Republic of Senegal.)

These are common species on sandy soils. On sandy clay soils or clay soils, *Andropogon amplectens*, *Schoenefeldia gracilis*, *Sporobolus festivus* and *Tetrapogon spathaceus* are more characteristic. *Aeluropus littoralis* is common along the shore of Mauritania, and in the delta of Senegal *Sporobolus marginatus* and *S. helvolus* are frequent. This is the grass cover *par excellence* for cattle production as all grasses are palatable and readily grazed and only some of the woody stems of *Aristida* are not eaten. It is mostly associated with scattered trees of *Acacia* and *Commiphora*. It is difficult to decide in which direction succession might progress with reduction of pressure of the biotic factors—possibly in the east towards the predominantly Saharo-Sindian species, *Cenchrus ciliaris* and *C. setigerus*, possibly in the west to the African grasses, species of *Andropogon*, *Hyparrhenia* and *Brachiaria*.

Transect borders in West Africa

South of the transect, there are some major types of grass cover characterized (from north to south) by species of *Andropogon*, *Hyparrhenia* and *Pennisetum*. Examples of some non-transect grass communities are given below, from countries to the south, such as Haute Volta, Niger, Dahomey, Ghana and Uganda.

H G. Kmoch[123] has given particulars of some vegetation zones and their component grass species (all African rather than monsoonal) which he recognizes in Upper Volta, on the basis of a study of a relatively restricted area on the ground and a comparison with aerial photographs of the whole territory taken by l'Institut géographique national de la France, in 1952.

(*a*) Gallery forest, no grasses.

(*b*) Plain of *Scirpus* and *Vetiveria* (*nigritana*) and *Panicum maximum* flooded for some weeks in the rainy season.

(*c*) Marshy areas with Cyperaceae, *Leersia* and *Paspalum scrobiculatum*.

(*d*) Wooded savannah, with *Andropogon gayanus* (especially its subspecies *bisquamulatus*), *Cymbopogon giganteus*, *Hyparrhenia sulcata* and other species of *Hyparrhenia*. Abundance of large trees exceeding 15 m. in height: *Afzelia africana*, *Isoberlinia doka*, *Khaya senegalensis*, *Pterocarpus erinaceus*, *Terminalia macroptera*, *Burkea africana*, *Sterculia setigera*, *Tamarindus indica*. In this and the three succeeding zones, shrubby species of *Combretum* and *Acacia* progressively increase from low frequency to higher frequency, with the taller trees gradually disappearing.

(*e*) Wooded savannah, with *Andropogon pseudapricus*. On the poorer soils this species replaces *A. gayanus* and *Hyparrhenia sulcata*, and is itself replaced by *Andropogon amplectens* on still less favourable sites.

(*f*) Wooded savannahs with *Andropogon amplectens*.
(*g*) Savannah with *Loudetia* and *Ctenium* spp.

The grass communities in the woodland savannah (*Acacia tortilis, A. seyal, Zizyphus lotus*) at the experimental farm at Filingué, near Niamey, Niger, appear to be a mixture of both Saharo-Sindian and African:[162]

Grasses:

Aristida mutabilis
Brachiaria distichophylla
Cenchrus ciliaris
C. leptocanthus
Chloris prieurii
Dactyloctenium aegyptium
Digitaria gayana
Echinochloa colonum
E. stagnina
Eragrostis ciliaris
E. senegalensis
E. tenuiflora
Hyparrhenia raynechtii
Panicum laetum
Pennisetum pedicellatum
Schizachyrium exile
Schoenefeldia gracilis
Urochloa insculpta

The associated legumes are:

Alysicarpus vaginalis
Crotalaria podocarpa
Tephrosia leptostachya
Zornia diphylla

At Okpara Farm in Dahomey,[187] however, the grass species are again almost entirely African:

Andropogon gayanus
A. pseudapricus
A. schirensis
Beckeropsis uniseta
Ctenium elegans
C. nubicum
Hyparrhenia chrysargyrea
H. dissoluta
H. confinis
Monocymbium ceresiiforme
Pennisetum pedicellatum
P. polystachyon
Schizachyrium sp.

In humid places:

Andropogon sp.
Setaria sphacelata
Tristachya sp.
Vetiveria nigritana

On the humus-rich banks of rivulets:

Andropogon sp.
Hyparrhenia diplandra
H. rufa
Panicum maximum
Schizachyrium cf. *platyphyllum*
Urelytrum sp.

In areas with an apparent lateritic layer:

Loudetia cf. *simplex*
Tristachya sp.

The probable type of subclimax and climatic climax vegetation in the northern Guinea Savannah Zone in the Farafara area of northern Ghana has been discussed by J. M. Ramsay and R. Rose Innes.[181] The grass cover of the degraded Guinea Savannah vegetation on which studies on burning were made may be seen from Table 7.

Another interesting type (CE.2) recognized by Rattray in Uganda is dominated by *Cenchrus ciliaris* in association with a number of Saharo-Sindian and African species: *Cenchrus ciliaris—Heteropogon contortus—Pennisetum ramosum*. (Other common grasses are *Bothriochloa pertusa*, *B. radicans*, *Brachiaria* spp., *Chloris roxburghiana*, *Chrysopogon aucheri*, *Cynodon* spp., *Enneapogon cenchroides*, *Panicum* spp., *Sporobolus* spp. and many others.)

This type occurs in a climatic/soil/topographic complex with types S.1 and H.27. On poorly drained clays, *Setaria incrassata* becomes dominant (S.1); *Hyparrhenia filipendula* and *H. dissoluta* (H.27) are co-dominant over the most mesophytic areas within the complex. It is associated with a savannah of mixed *Acacia* spp. and deciduous broad-leaved trees. It occurs at altitudes of 760–1,220 m. with a rainfall of 630–1,020 mm., over 90 per cent falling in the months March to October inclusive. It has a carrying capacity of about 1 adult bovine (Boran type Zebu) to 4 hectares.

SUDAN

We must distinguish between the Sudanian zone in West Africa and the political entity of the Sudan. For this territory we have M. N. Harrison's[82] account of the soils and types of vegetation.

Vegetation	Approx. area thousand sq. kilometres	Rainfall mm.
I Desert	730	less than 75
II Semi-desert	494	75–300
III Woodland savannah	1,000	300–1,500
IV Flood region	247	
V Montane vegetation	6.5	

14. Upper Volta, Guinea Zone, region Bobo-Gaova. *Hyparrhenia rufa*. (*Photo H. G. Kmoch*)

15. Upper Volta, Sahel Zone, region Dori. *Schoenefeldia*, *Ctenium*, *Cenchrus*. (*Photo H. G. Kmoch*)

16. Upper Volta. Plain of *Scirpus*, *Vetiveria*, *Panicum maximum*, Samandeni (?). (*Photo H. G. Kmoch*)

17. Upper Volta. Savannah forest with *Andropogon gayanus*, Samandeni. (*Photo H. G. Kmoch*)

18. Upper Volta. Savannah forest with *Andropogon pseudapricus*, Samandeni. (*Photo H. G. Kmoch*)

19. Upper Volta. Savannah forest with *Andropogon amplectens*, Samandeni. (*Photo H. G. Kmoch*)

20. Upper Volta. Savannah forest with *Loudetia* and *Ctenium* Samandeni. (*Photo H. G. Kmoch*)

TABLE 7

Summarized Grass Cover and Composition Analyses in northern Ghana[181]

(Means of six transects totalling 1,200 points in each plot)

Species	I Late Burn		III Early Burn		II Protected	
	Hits	%	Hits	%	Hits	%
Loudetia acuminata	90	59·6	59	38·8	49	43·8
Andropogon pseudapricus	19	12·6	26	17·1	8	7·1
Schizachyrium schweinfurthii	12	7·9	7	4·6	2	1·8
Sporobolus festivus	5	3·3	1	0·7	—	—
Andropogon amplectens	3	2·0	5	3·2	—	—
Aristida kerstingii	3	2·0	2	1·3	—	—
Tripogon minimus	3	2·0	2	1·3	—	—
Dicotyledon spp.	3	2·0	—	—	4	3·6
Andropogon gayanus var. *bisquamulatus*	2	1·3	14	9·2	28	25·0
Andropogon schirensis	2	1·3	7	4·6	2	1·8
Cymbopogon giganteus	2	1·3	6	3·9	10	8·9
Schizachyrium nodulosum	2	1·3	2	1·3	—	—
Monocotyledon spp.	2	1·3	2	1·3	3	2·7
Brachiaria sp.	1	0·7	—	—	—	—
Digitaria sp.	1	0·7	—	—	—	—
Sedge spp.	1	0·7	1	0·7	—	—
Schizachyrium sanguineum	—	—	11	7·2	—	—
Hyparrhenia chrysargyrea	—	—	3	2·0	—	—
Hyparrhenia sp.	—	—	2	1·3	1	0·9
Diectomis fastigiata	—	—	1	0·7	—	—
Hyparrhenia notolasia	—	—	1	0·7	—	—
Rottboellia exaltata	—	—	—	—	3	2·7
Grass sp.	—	—	—	—	2	1·8
Totals	151	100·0	152	99·9	112	100·1
Total Basal Cover (%)	12·6		12·7		9·3	
Extreme variation for 6 transects	8·0–17·5%		9·0–16·5%		7·0–125%	
No. of species encountered	16		18		11	

Soils:

(*a*) various desert soils
(*b*) stabilized dune sand or 'Qoz'
(*c*) dark cracking clays, including flood region soils
(*d*) non-cracking clays
(*e*) laterite catena soils
(*f*) various hill soils.

One gains the impression in reading of the gramineous species found within each type of vegetation that the grass covers and their associated tree species of the semi-desert and of the low-rainfall woodland savannah belong to the monsoonal transect. Part of the low-rainfall woodland savannah and all the high-rainfall woodland savannah are tending towards more typical African grasslands containing species of *Hyparrhenia*, *Andropogon* and *Brachiaria*.

Before summarizing Harrison's work, it is appropriate to give particulars of two *Aristida* types and one *Cenchrus biflorus* type which Rattray[183] recognizes as important grass covers in our general transect within Africa.

A.11 SUDAN

Aristida spp. (*A. papposa*, *A. funiculata*, *A. mutabilis*, *A. hirtigluma* and *A. plumosa* are common species)—*Panicum turgidum* —*Schoenefeldia gracilis*—*Lasiurus hirsutus*

Panicum turgidum is a characteristic perennial species in this type of grass cover. *Cenchrus setigerus* becomes abundant in the east on some sandy soils, while *Sehima ischaemoides* and the grasses of the drier parts of type Sorghum.1 come in on clay soils in the wetter parts. *Cymbopogon proximus* is a good indicator of overgrazing throughout the region. Areas of *Cyperus conglomeratus* also occur on sand. It is a sparse, semi-desert, predominantly annual grass cover, usually associated with a tree or bush steppe of *Acacia* species that vary much with soil type. On mixed soils there is *Acacia tortilis*/*Maerua crassifolia* scrub: on clay soil, open grassland with only occasional patches of *Acacia ehrenbergiana*; on sandy soils a varying density of *Acacia* spp. (*A. tortilis* subsp. *raddiana*, *A. mellifera* and *A. tortilis*) with *Commiphora* spp. and by the Red Sea coast a scrub of *Acacia asak* and *A. etbaica*. This type occurs from the coast up to an altitude of 1,200 m. with a rainfall range of 80–300 mm. on sand and 380 mm. on clay, falling in a season of only two or three months from July to September.

A.12 SUDAN
Aristida spp. (*A. pungens, A. plumosa* are common)—*Panicum turgidum* (in 'wadis')

This is the eastern part of the Libyan Desert where vegetation is virtually absent except along watercourses, or occurs as ephemeral herbs and grasses springing up after rare showers. It occurs from the coast up to an altitude of 800 m. with a rainfall varying from nil to 900 mm. falling predominantly in August. In two areas a few nomads with their camels and goats lead a precarious existence on vegetation by watercourses chiefly of the *Acacia tortilis/Maerua crassifolia* desert scrub type.

Rattray's type based upon *Cenchrus biflorus* is:

CE.4 SUDAN
Cenchrus biflorus—C. ciliaris—Eragrostis tremula—Aristida spp. (*A. pallida* and *A. sieberiana* are common)

The *Cenchrus* and most of the *Eragrostis* are secondary after cultivation. In the drier parts with some clay admixture are areas of *Aristida mutabilis* and *Schoenefeldia gracilis*. Increasingly in the wetter parts are *Ctenium elegans, Loudetia hordeiformis* and *Andropogon gayanus*. In certain localities, where the sand alternates with clay flats, *Schoenefeldia* and annual *Aristida* spp. (including *A. funiculata, A. submucronata* and *A. mutabilis*) with an admixture of *Sporobolus marginatus, Dactyloctenium aegyptium, Brachiaria* sp. (aff. *B. xantholeuca*), *Chloris pilosa* and *Ch. prieurii* are found. The predominantly annual grass cover is associated with *Acacia senegal* savannah in the drier parts, and in the wetter parts with a mixed savannah of *Combretum cordofanum, Dalbergia melanoxylon, Albizzia amara* subsp. *seriocephala, Guiera senegalensis, Sclerocarya birrea* and *Terminalia* spp. (*T. laxiflora* and *T. brownii*). It occurs at an altitude of 450–1,100 m. with a rainfall range of 300–640 mm. falling in 4 months, June to September on 'qoz' sands (fixed sand dunes).

TABLE

Ecoclimatic gradient in eastern

Zones	Vegetation (Keay)[4]	Forest formations	Chief tree species	Grass cover types (Rattray)[183]	
Eastern Equatorial desert and sub-desert	Desert. Sub-desert steppe; tropical types	—	—	*Chrysopogon* (5); desert conditions	
Eastern Equatorial savannah	Wooded steppe with abundant *Acacia* and *Commiphora*	*Acacia–Commiphora* woodland (Sahelian Zone in West Africa)	*Acacia* *Commiphora* *Dobera* *Salvadora* *Boscia* *Phyllanthus* *Euphorbia*	*Chloris* (1) ass. with tree steppe *Chrysopogon* (1, 2, 3, 4, 6, 6A) ass. with tree steppe	
Eastern Coastal	Coastal forest–savannah mosaic			*Panicum* (1, 2) ass. with woodland *Hyparrhenia* (5) ass. with woodland	
Lake Victoria	Forest–savannah mosaic			*Eragrostis* (9) ass. with savannah *Hyparrhenia* (25, 26, 27) ass. with savannah *Hyparrhenia* (23) ass. with woodland	*Panicum* (2, 3) ass. with savannah *Themeda* (12, 15) ass. with savannah
Eastern and Central Plateau	*a.* Woodlands of various types with *Brachystegia* and *Julbernardia*, etc.	Miombo and Mopane woodlands	*Brachystegia* *Isoberlinia* *Julbernardia* *Oxytenanthera* *Pterocarpus* *Colophospermum*	*Aristida* (4) ass. with woodland *Aristida* (5, 6, 7) ass. with savannah *Eragrostis* (6, 7, 8) ass. with savannah *Heteropogon* (1) ass. with savannah	*Loudetia* (2) ass. with grassland *Loudetia* (1, 3, 4) ass. with savannah *Loudetia* (5, 6, 7) ass. with savannah *Pennisetum* (1, 2); grassland
	b. Acacia, Combretum and *Terminalia* in savannah woodlands			*Hyparrhenia* (3, 5, 6, 10, 15, 17, 23) ass. with woodland *Hyparrhenia* (13, 14, 17B, 24); grassland	*Themeda* (12, 13) ass. with savannah *Themeda* (8, 9, 10, 11); grassland
	c. Cryptosepalum pseudotaxus, Baikiaea, Marquesia, and *Guibourtia coleosperma* on Kalahari sands, and			*Hyparrhenia* (4, 5A, 9, 12, 15A, 16, 17A, 19, 21); ass. with savannah	
	d. Colophospermum mopane abundant				
East African Highlands	Montane grassland and forest			*Heteropogon* (2) ass. with savannah *Hyparrhenia* (30) ass. with savannah *Pennisetum* (1, 2) ass. with woodland	*Pennisetum* (3) ass. with savannah *Setaria* (3) ass. with woodland *Themeda* (16) ass. with savannah

nd central regions of Africa[253]

ain types of land-use	Main crops	Livestock	Countries	Zones
omadic grazing	—	Camels, sheep and goats–some cattle	Somalia, Ethiopia	Eastern Equatorial desert and sub-desert
		Cattle, sheep and goats	N. Territory of Kenya	
omadic grazing and culti- tion generally subsidiary herding. Minor tree oducts and protection ainst desiccation, wild- e	Millet, sorghum	Cattle, sheep and goats	Somalia, Ethiopia, Sudan, Kenya, Tanzania	Eastern Equatorial savannah
rable cultivation with rying periods of fallow	Rice, cassava, coconut	Few cattle and goats	Somalia, Kenya, Tanzania, Mozambique	Eastern Coastal
razing and arable culti- tion with variable but nerally short fallow riods. Tree crops ainly coffee	Plantains, bananas, finger millet, sorghum, robusta coffee, cotton	Cattle and goats	Kenya, Uganda, Tanzania	Lake Victoria
Arable cultivation with variable but generally long fallow periods. European farming and tobacco planting	*a.* Finger millet, sorghum, cassava, maize, flue-cured and Turkish tobacco. Limited sown pastures in European farming	*a.* Few livestock of all classes– limited by tsetse	Tanzania, Congo (Kinshasa), Angola, Zambia, Rhodesia, Malawi, Mozambique	Eastern and Central Plateau
Grazing and arable cultivation with variable but generally short fallow periods. European maize, cattle and mixed farming	*b.* Maize, sorghum, sown pastures in European farming	*b.* Cattle with some sheep and goats depending on absence of tsetse		
Grazing and arable cultivation with variable but generally long fallow periods on the sands, and short fallow periods and permanent cultivation on associated alluvial and peat soils	*c.* Bulrush millet, cassava, maize, finger millet	*c.* Cattle with some sheep and goats depending on absence of tsetse		
Ranching and grazing in tsetse-free areas. Arable cultivation mainly on associated alluvial and other soils	*d.* Maize, sorghum, bulrush millet	*d.* Cattle with some sheep and goats		
razing and arable culti- tion with short fallow riods and permanent ltivation of bananas d plantains in moist rest zones. European ttle ranching, mixed rming and coffee anting	Maize, sorghum, plantains and bananas, coffee, pyrethrum, tea, limited sown pasture	Cattle predominant, goats and some sheep	Ethiopia, Kenya, Rwanda, Burundi, Tanzania, Congo Kinshasa, Malawi, Rhodesia	East African Highlands
oodland utilization for cal consumption. Forest antations				

Vegetation of Sudan[82]

I. DESERT

There is vegetation only in watercourses but sufficient to support a thin scatter of hardy nomads, and the ephemeral herbs and grasses that spring up after rare rain showers and give rise to valuable pasturage. The best grazing species are *Indigofera bracteolata*, *I. arenaria* and *Neurada procumbens*.

II. SEMI-DESERT

subdivision	sq. kilometres
(*a*) *Acacia tortilis/Maerua crassifolia* desert scrub	187,000
(*b*) Semi-desert grassland on clay	104,000
(*c*) Semi-desert grassland on sand	86,000
(*d*) *Acacia mellifera/Commiphora* desert scrub	86,000
(*e*) *Acacia glaucophylla/A. etbaica* scrub	31,000
	494,000

This differentiation, in particular the proportion of scrub bushes to grass and herbs, depends more on soil type than on rainfall. The reasons for the transition from (*a*) to (*d*) between longitudes 30° and 32° N are not known but may be due to a number of causes. Geological erosion, far advanced on various desert soils, gives rise to very varied vegetation. Sharp separation into hard-surfaced off-flow soils completely bare of vegetation at one extreme, and on-flow soils that have scrub bushes and even some trees at the other extreme; gradual transition on one side to desert conditions; under higher rainfall, scrub bushes increase to thickish more uniform stands. Where soils are not greatly eroded, they easily absorb all the scanty rain that falls on them—resultant vegetation is an even cover of mixed grasses and herbs called semi-desert grassland. Scrub bushes occur only where the surface creates water-receiving sites.

Subdivision (*a*) has much variety and is used as a blanket classification into which several other types of vegetation are included. The various desert soils of the west bearing subdivision (*d*) alternate with about equal areas of sand soils bearing semi-desert grassland—dividing line between the two subdivisions arbitrary. The west has not been subject to so much erosion as east—it is conceivable that the mixture in the west of subdivisions (*c*) and (*d*) is a stage of transition towards (*a*), and that the conditions of the east will be reached 'after many ages of further erosion'.

The best grazing plants in the western part of the semi-desert (*c* and *d*) are *Aristida plumosa*, *Blepharis linariifolia* (?) and *Monsonia* sp. When these species are eliminated by overgrazing, they are replaced by *Schmidtia pappophoroides*, *Cenchrus biflorus*, *Eragrostis tremula* and annual species of *Aristida*. During rains ephemeral legumes improve grazing and may cause bloat. Ten animal units per square mile are optimal for both subdivisions. The best browse plants are *Acacia raddiana*, *A. mellifera* and *Maerua crassifolia*. On *Acacia tortilis/Maerua crassifolia* Desert Scrub, only one of the three most abundant perennial grasses, *Panicum turgidum*, is of any value for grazing, while most of the annuals are poor. Browse bushes and herbs therefore provide much of the animal feed.

On Semi-Desert Grassland on Clay, a species of *Blepharis* (? *B. edulis*) is the best dry season grazing for camels and sheep. The most abundant grasses are practically useless for grazing, apart from *Sehima ischaemoides* and the annuals of non-cracking clays, *Schoenefeldia gracilis* and annual species of *Aristida*, but for cattle rather than for camels and sheep. Carrying capacity is seven animal units per square mile.

III. WOODLAND SAVANNAH

	Major subdivisions	sq. kilometres
A1.	Low rainfall woodland savannah on clay	317,000
A2.	Low rainfall woodland savannah on sand	216,000
	Special areas of low rainfall woodland savannah	159,000
B.	High rainfall woodland savannah	348,000
		1,040,000

Woodland Savannah covers the largest part of the Sudan—there is a wide variety of distinct types of vegetation under widely differing conditions. Separations into major and minor subdivisions are made on the basis of character of vegetation and species composition, but major subdivisions are named by the rainfall and soil variations with which they are associated.

The separation between low and high rainfall comes between 800 and 1,000 mm.

Low rainfall woodland savannah

Trees in drier parts are nearly all low thorn, principally *Acacia* spp. with some thorn bushes and shrubs including thickets of *A. mellifera*. Broad-leaved deciduous trees become predominant in wetter parts, but usually with a proportion of thorn. There are more annual grasses than perennials (proportion of perennials less than in parts of the Semi-Desert)—perennials seldom attain the height and coarseness of perennial grasses in high rainfall woodland savannah. A variable proportion of herbs is always found.

The central belt of the Sudan (LRWS belt) has no soils formed *in situ*, other than localized hill soils. It is sharply divided into two extreme soil types—Dark Cracking Clays of the east and Stabilized Sand Dunes or Qozes of the west. The major subdivisions recognized on clay and sand are each further subdivided into three different rainfall belts. On the uniform plains of Dark Cracking Clay, the transition from one rainfall belt to the other is sharply defined by change in tree dominants, although grasses show no good line of demarcation. On the Stabilized Sand Dunes, with variation due to type of sand and position on slope, and with a larger variety of tree species, the three rainfall belts are not clearly or sharply separated but merge gradually. The proportions of the different species found rather than their presence and absence have to be taken as the distinguishing criteria.

The grasses of LRWS on Clay often occur in separate patches with each patch dominated by a single grass almost to the exclusion of others. The marked changes from one patch to the other take place suddenly and for no apparent reason. *Acacia mellifera* thickets and the drier parts of *A. seyal/Balanites* savannah alternate in time and space with grassland areas in what has been described by J. Smith in the Sudan Soil Conservation Committee Report, 1944, as the grassland/*Acacia* cycle.

Also in LRWS are four small, very different areas of mixed vegetation owing their existence to local conditions; these are lumped together and reported in a category—Special Areas of LRWS.

From the grazing point of view all the western part of LRWS

is dealt with as one because the different parts are covered by the Baggara tribes during their seasonal migrations. The most important grazing plants of the two northern subdivisions, *Acacia senegal* savannah and *Combretum cordofanum/Dalbergia/Albizzia seriocephala*, are *Eragrostis tremula*, *Cenchrus biflorus*, *Blepharis* (?) *linariifolia*, *Zornia diphylla* and annual species of *Aristida*. *Cenchrus biflorus* is palatable in the rains. *Blepharis* is the chief grazing for camels and sheep, but useless for cattle in dry season—all they have is poor *Eragrostis* and *Aristida*.

The dry season regrowth after burning of the southern subdivision, *Terminalia/Sclerocarya/Anogeissus/Prosopis* savannah is mainly *Andropogon gayanus* var. *bisquamulatus*. Regrowth is very unreliable—better if not burnt. Where water is sufficient and transport facilities permit, there is excessive overcultivation, as in the 'Kordofan graveyard' along the railway line. Here the sand is blowing again and the dominant feature of the vegetation in the worst areas is *Calotropis procera*.

On the Hill Catena, which is largely cultivated, the grazing is poor except in the wadis of the western Darfur Hill Catena, where there are broad belts of *Acacia albida* and *Cynodon dactylon*. The tree drops clumps of ripe pods in the dry season, giving a yield as high as any bean crop that might be grown as a protein supplement to the already good grazing on *Cynodon dactylon*. The tribe that uses this grazing has large cattle which suffer little loss from starvation in bad years.

The Baggara Repeated Pattern grazing, in particular its saltiness, is of outstanding value. The three best and 'saltiest' grasses occurring most abundantly in the intermediate zone between 'atamur' and 'nagaa' are *Sporobolus marginatus*, *Dactyloctenium aegyptium* and *Brachiaria* sp. aff. *xantholeuca*. Most of the surface of the 'nagaa' has only a scanty cover of the poor grazing annual *Aristida* spp. and *Schoenefeldia gracilis*.

Apart from the different *Acacia* trees, whose usefulness is far less than their abundance, the best browse bushes are *Cadaba afrinosa*, *Grewia betulaefolia*, *Dichrostachys glomerata*, *Bauhinia rufescens* and *Balanites aegyptiaca*.

The optimal carrying capacity of all the western part of the LRWS is about 30 animal units per square mile.

On Low Rainfall Woodland Savannah on Clay, all three subdivisions are covered by migrations of tribes which use the grazing. Most of the grasses of the typical clay plains are 'harig' grasses—harig cultivation involves burning off an old stand of grass before sowing a crop. The grasses suitable for harig cultivation are *Sorghum purpureo-sericeum*, *Hyparrhenia pseudocymbaria*, *Cymbopogon nervatus*, *Brachiaria obtusiflora* and *Setaria incrassata*. Only *Brachiaria* provides good dry season grazing but is limited to small low-lying areas of black clay. The others are practically useless in the dry season—animals starve rather than eat them. Associated with *Acacia mellifera* in the north are good short annuals such as *Sehima ischaemoides*, *Tetrapogon spathaceus* and *Setaria verticillata*. In *Anogeissus/Combretum hartmannianum* Savannah Woodland, particularly in low-lying areas, there is some regrowth after burning of *Hyparrhenia rufa*, *Andropogon gayanus* var. *squamulatus* and *Ischaemum brachyatherum*. The present stocking rate for typical (harig) and atypical areas is about 15 animal units per square mile but that should be reduced by about 20 per cent to stop deaths from starvation.

High rainfall woodland savannah

There are broad-leaved trees of many species, varying according to site; trees taller than in LRWS—few thorn trees and very few bushes. Perennial grasses predominate, growing in coarse tall tussocks, often widely separated. After burning in early dry season the grasses give a short green regrowth—fires fiercer than in LRWS because of greater bulk. HRWS occurs on Laterite Catena Soils.

Above 1,300 mm. rainfall, the climax would be rain forest, as witnessed by relict patches of forest, relict forest trees in savannah woodland, ease with which forest species recolonize the savannah as soon as fire is excluded. A narrow strip along the Congo and Uganda borders is distinguished as Woodland recently derived from Rain Forest.

The main grasses of HRWS are the coarse tall tussocks of *Hyparrhenia* spp. and *Andropogon gayanus*, little use for grazing except in the early rains and for their limited regrowth after burning in dry season. When trypanosomiasis is eliminated, the land settled, grass grazed down and fires prevented, these grasses will become replaced by useful pasture species such as *Setaria sphacelata*, *Brachiaria brizantha*, *Chloris gayana*, *Panicum gayanum* and *P. coloratum*.

IV. FLOOD REGION

Classification given in Equatorial Nile Project Report, 1954.

High Land

Forest

(*a*) Palm type

(*b*) Poorly developed deciduous broad-leaved type

(*c*) Mixed *Acacia*-dominant type

(*d*) *Acacia seyal/Balanites aegyptiaca* type

Grassland

Intermediate Land

Forest

Acacia seyal/Balanites aegyptiaca type

Grassland

(*a*) *Hyparrhenia rufa* type

(*b*) *Setaria incrassata* type

(often as pure stands but may have associated small amounts of *Andropogon gayanus* var. *squamulatus*, *Ischaemum brachyatherum*, *Cymbopogon giganteus*)

Swamp

Seasonal swamp

(*a*) Inland Khor-bed type

(*b*) Riverain type

Permanent papyrus swamp

(associated with various climbers, miscellaneous species and *Leersia hexandra*).

Important grasses in the Seasonal Swamp are *Echinochloa pyramidalis* and *Oryza barthii*, which are burnt for their regrowth, like *Hyparrhenia rufa* and *Setaria incrassata* on the Intermediate Land. *Echinochloa stagnina*, the best Seasonal Swamp grazing grass, is very palatable in its mature state—it is not burnt but produces a regrowth after it has been grazed down. *Vossia cuspidata* can also be grazed as it stands. Cattle in the Flood Region which are kept on these grasses during the dry season thrive better than they do in the rains. The dry season carrying capacity of *Echinochloa pyramidalis* is about 2 feddans per beast, of *E. stagnina* about 1 feddan per beast.

V. MONTANE VEGETATION

Little in common other than the fact that they owe their differences from the surrounding vegetation to the altitude and rainfall of the mountain masses on which they occur.

Areas of Montane Vegetation are:

(*a*) The Imatong and Dongotona Mountains
(*b*) The Didinga Mountains
(*c*) The Red Sea Hills
(*d*) Jebel Marra.

NORTH AFRICA

Here we find winter-rainfall conditions along a relatively narrow coastal strip, in complete contrast to the summer rainfall or false monsoonal conditions on the other side of the Sahara. Two of Rattray's[183] *Aristida* types (A.16 and A.17) extend far north towards the Mediterranean, and from Morocco in the west to Egypt in the east.

A.16 MOROCCO, ALGERIA, TUNISIA, LIBYA, EGYPT
Aristida plumosa—A. obtusa—A. acutiflora—A. ciliata (a very palatable species)—*A. pungens* (on the dunes)
(In Libya, *Stipa lagascae*, *S. capensis* and *S. barbata* have also been recorded from this zone.)

These grasses are associated with a steppe type of vegetation consisting of small scattered trees of *Acacia* and dwarf shrubs. *Stipa* sp., *Genista* and *Artemisia* are often found in the 'wadis'. *Lygeum spartum* is common on gypseous soils but not on desert sands. It occurs at altitudes from sea level to 1,000 m. with a winter rainfall of less than 200 mm.

A.17 MOROCCO, ALGERIA, TUNISIA, LIBYA, EGYPT
Aristida spp.—*Panicum turgidum—Lasiurus hirsutus—Pennisetum divisum*
(The last three species are almost confined to the 'wadis'.)

These grasses are characteristically found in the Sahara, associated at times with scattered *Acacia* species and shrubs such as *Anabasis*, *Haloxylon* and *Cornulaca*. Extensive areas, however, are devoid of vegetation and only an ephemeral cover of grass and herbs appears after rain. The rainfall is less than 500 mm. annually.

Rattray's Mediterranean types proper are:

H.39 TUNISIA, LIBYA, EGYPT
Hyparrhenia hirta—Oryzopsis miliacea—Cynodon dactylon

Ph.1 MOROCCO, ALGERIA, TUNISIA
Phalaris minor—Ph. brachystachys—Ph. paradoxa—Lolium rigidum—Oryzopsis miliacea—Hordeum bulbosum

Ph.2 LIBYA
Phalaris spp.—*Hordeum bulbosum—Poa bulbosa—Lolium rigidum—Oryzopsis miliacea—Dactylis glomerata*

St.1 MOROCCO, ALGERIA, TUNISIA, LIBYA
Stipa tenacissima—Lygeum spartum (on gypsum soils)

St.2 MOROCCO, ALGERIA
Stipa capensis—in almost pure stands, particularly in the south.

St.3 TUNISIA, LIBYA
Stipa lagascae—S. parviflora (common on overgrazed areas and shallow soils).

UNITED ARAB REPUBLIC (EGYPT)

M. Kassas of the Fouad I University in Cairo has made between 1952 and 1965 a series of fundamental studies on habitat and plant communities in the Egyptian desert.[103, 104, 110, 111, 113, 114] The objective was stated in 1952 'to formulate a schema for the mechanism of vegetational dynamics in the desert'. The cycle of arid erosion includes youth, mature and old-age stages; the plant cover should be most responsive to such changes. It was therefore hoped to show the intimate relation between the changing desert surface and the plant cover. By a study of habitats and plant communities on rocky plateaux, desert wadis, desert mountains, gravel desert and desert plains, it has been possible in the last paper to define the community types within one of the principal desert ecosystems, the wadi beds. The successional relationships are shown in Table 9.

Pennisetum dichotomum is dominant where soil is mainly calcareous silt. *Panicum turgidum* and *Desmostachya bipinnata* may be dominant when soil is a mixture of alluvial calcareous silt and sand. *Panicum turgidum* is a desert grass with a wide range of distribution, especially on sand drifts and low sand dunes. *Desmostachya bipinnata* belongs more to the Nile valley than to these desert wadis. The grassland type marks a turning point in the structure of the wadi-bed vegetation from annual to perennial dominants.

In the study on the gravel desert a community on affluent runnels is dominated by *Lasiurus hirsutus*, and one in the main channels by *Panicum turgidum*.

TABLE 9

Successional relations between the community types in a desert ecosystem, in wadi-beds[111]

Stage				
CLIMAX STAGE—deep wadi-fill, underground	Type XIII *Tamarix mannifera*			
FIFTH STAGE (scrubland)—deeper deposits	Type IX *Lycium arabicum*	Type X *Nitraria retusa*	Type XI *Atriplex halimus*	
FOURTH STAGE (grassland)—silt terraces	Type VIII *Pennisetum dichotomum*			
THIRD STAGE—deeper deposits	Type VII *Zilla spinosa*	Type VIa *Anabasis articulata*		
SECOND STAGE—shallow coarse deposits	Type II *Zygophyllum decumbens*	Type III *Zygophyllum coccineum*	Type IV *Zygophyllum album*	Type V *Anabasis setifera*
PIONEER STAGE (chasmophyte)—rock surface (no soil)	Type I *Stachys aegyptiaca*			

Further studies of the Red Sea coastal lands[114] have revealed *Aeluropus lagopoides* and *Sporobolus spicatus* in the salt marshes, and *Panicum turgidum*, *Cenchrus setigerus* and *Lasiurus hirsutus* on the plains. A *Juncus arabicus*/*Imperata cylindrica* type is noted where marshy habitats are produced by the accumulation of brackish water at the ground surface and its subsequent evaporation.

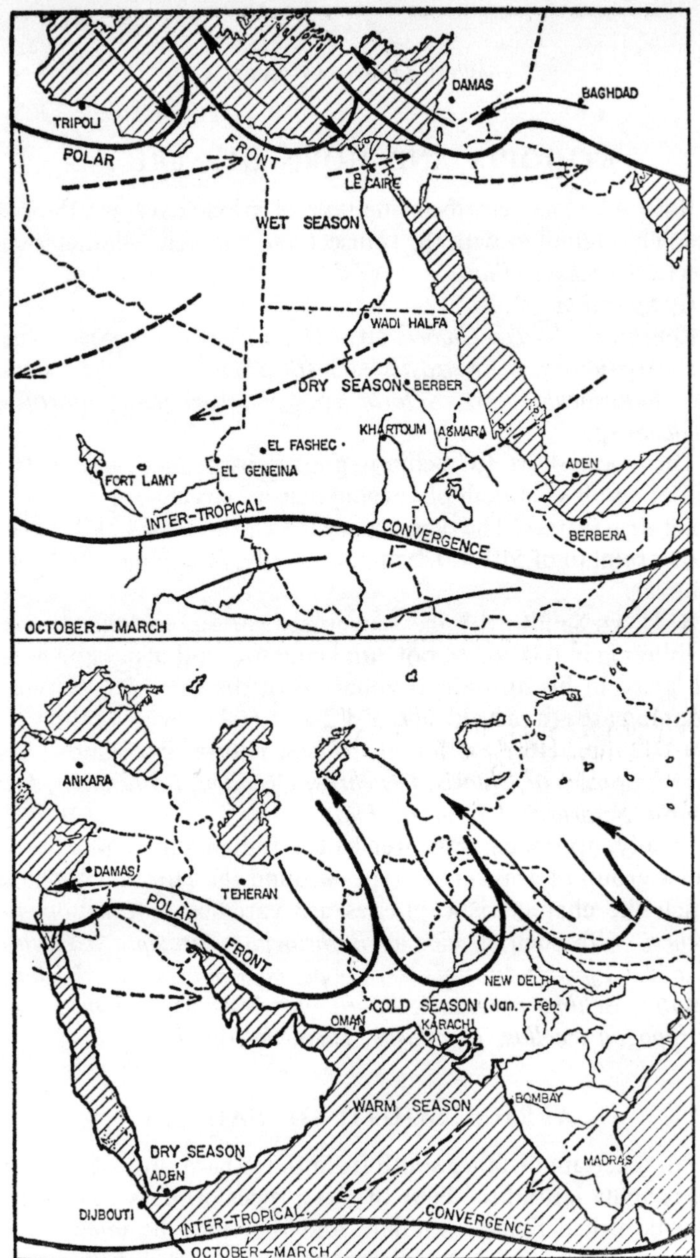

FIGURE 21

Polar front and Intertropical Convergence Zone in north-east Africa, Arabia and India (from an FAO document)

ETHIOPIA AND HORN OF AFRICA

Rattray[183] has described one type of grass cover in Ethiopia, which has affinities with the transect and has as its characteristic species *Cenchrus ciliaris.*

Composition:

Cenchrus ciliaris—*Chloris* sp.—*Hyparrhenia* sp.—*Sporobolus* sp.—*Aristida* sp.—*Eragrostis* sp.—*Brachiaria* sp.—*Pennisetum* sp.—*Bothriochloa* sp.—*Setaria* sp.—*Heteropogon contortus*—*Panicum* sp.

These are short to medium grasses associated with a tree and/or scrub savannah of varying density largely dominated by small *Acacia* trees. The type occurs at an altitude of 450–1,200 m. with a rainfall of 510–1,020 mm. annually, falling mostly during summer.

R. H. D. Sandford[195] has recognized various ecological zones in Ethiopia in relation to potential land use, and in passing notes the place in his altitudinal zonation of the semi-dry marginal cropping areas at an altitude of 1,200–1,600 m. with a rainfall of 450–600 mm. Here are found the best natural grasslands, containing species of *Chloris*, *Cenchrus*, *Cynodon*, *Pennisetum*, *Brachiaria*, *Setaria* and *Bothriochloa.*

Finally, approaching nearer to the Indian sub-continent, we find a group of grass cover types around the Horn of Africa in which the characteristic grasses are varieties of *Chrysopogon aucheri*, with such species as *Bothriochloa insculpta*, *Cenchrus ciliaris*, *C. pennisetiformis*, *Cymbopogon schoenanthus*, *Lasiurus hirsutus*, *Sehima nervosum*, *Sporobolus helvolus*, *S. marginatus*, *Tetrapogon tenellus*, *T. villosus*, *Themeda triandra.*

AFRICA SOUTH OF SAHARA

For a certain distance south from the Sahara proper, we have the roughly latitudinal belts of those types of grass covers that may be called Saharo-Sindian, and that may be regarded as belonging to the Afro-Asian-Australian transect of comparable grass communities that are the basis of the present proposition. At the eastern end in Africa, we find that two of our major species, *Lasiurus hirsutus* and *Panicum turgidum* extend up into the ecosystems of the Egyptian desert.[110] In other studies of the Red Sea coastal lands of Egypt,[108] we find reference to *Aeluropus lagopoides*, *Sporobolus spicatus*, *Panicum turgidum*, *Cenchrus setigerus* and *Lasiurus hirsutus*.

As already stated, the significance of the occurrence of major transect species far to the south of the Sudan, Ethiopia and Somalia on the eastern side of Africa is open to much discussion. Are they really indigenous there, or could they have travelled south with major human migrations? Are they truly monsoonal species that have found a summer-rainfall, high-altitude environment suited to their needs? This question has already been mentioned under *Plant dispersal and migration* in Chapter III. Also relevant, in addition to J. M. Rattray's map of the grass cover of Africa, are the papers by F. White[247] on the savannah woodlands of the Zambezian and Sudanian domains, De Winter[63] on the South African Stipeae and Aristideae, H. Wild[257] on additional evidence for the Africa/Madagascar/India/Ceylon land bridge, D. F. Vesey-Fitzgerald[237] on Central African grasslands, and J. K. Morton[153] on the composition, distribution and significance of the upland floras of West Africa in relation to climatic changes.

Chapter V

NEAR EAST

ISRAEL

Probably the most relevant work in the Near East in the present context is that of the Israeli botanist, A. Eig, who conducted his phytosociological studies according to the principles of the Montpellier school at the time of Braun-Blanquet. Eig worked from 1932 to 1938, and a posthumous summary of his conclusions was published in the *Palestine Journal of Botany*.[68] The author had classified and named the units of desert and steppe vegetation, but stated that his conclusions regarding the non-desert units were still tentative. Eig believed that the most important criteria are exact floristic study, knowledge of the geographical distribution and ecological requirements of the species and their forms, and the study of associations and their distribution. He advised against the use of climatic formulae.

It was concluded that Israel is a meeting point of the four plant geographical regions recognized by him (Fig. 22):

Mediterranean
Irano-Turanian
Saharo-Sindian
Sudano-Deccanian

One may presume that the third and fourth regions represent Near Eastern sections of our whole transect. It will be seen by comparing the descriptions of the communities in West Africa and West Pakistan and Western India that there is sufficient justification for the linking term, Saharo-Sindian. The term Sudano-Deccanian does not seem to be so firmly based.

The genera and species noted by Eig in those parts of Israel which come into these categories are:

Saharo-Sindian: *Salsola villosa, Stipa tortilis, Anabasis articulata, Notocerus bicornis, Gymnocarpus fruticosus, Zilla spi-*

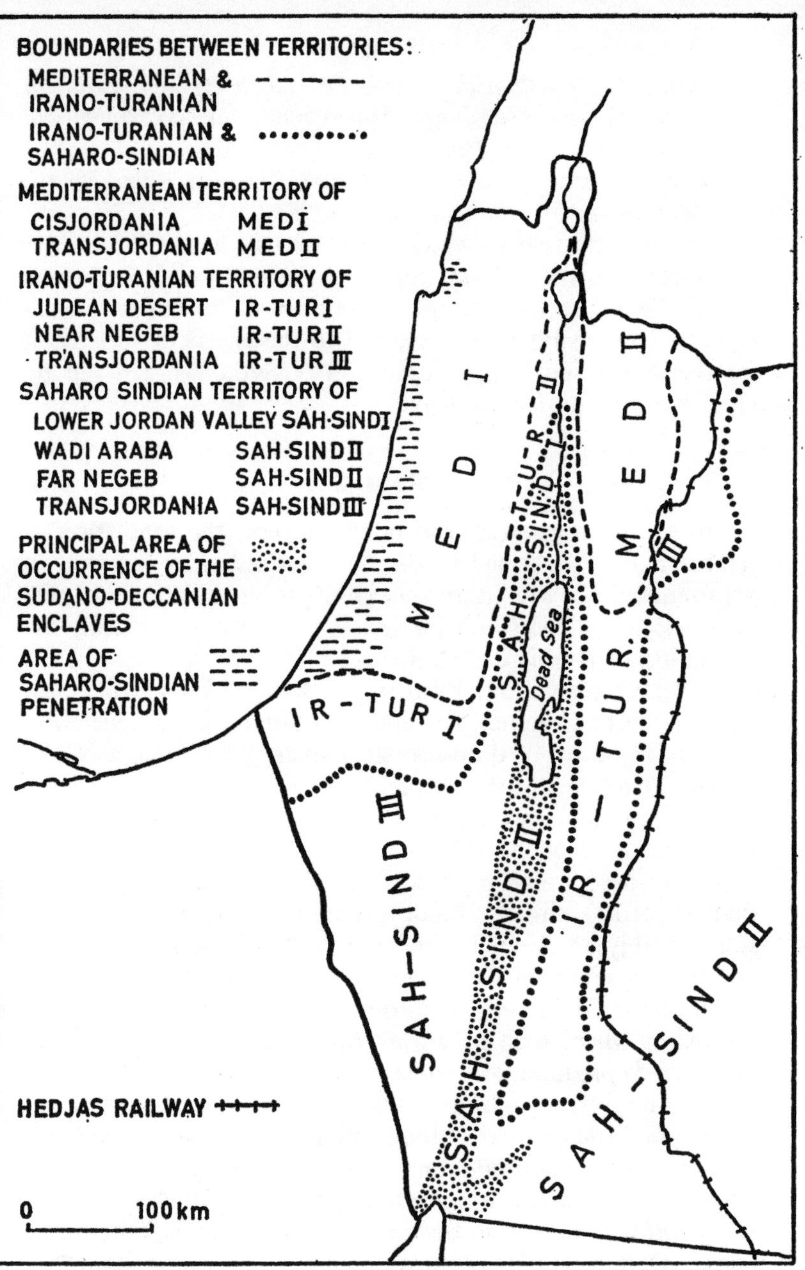

FIGURE 22

Plant geographical map of Palestine (Israel), according to A. Eig[67] *nosa, Zygophyllum dumosum, Chenoleia arabica, Erodium glaucophyllum, Herniaria hemistemon, Suaeda asphaltica, Atriplex palestina, Halogeton alopecurus*

Sudano-Deccanian: *Acacia tortilis, Zizyphus spina-christi, Moringa aptera, Balanites aegyptiaca*

The genera and species which are characteristic of the border zones between these major regions are also noted. Reference should be made to subsequent writings of botanists at the Hebrew University, Jerusalem, and to the charts of succession for the desert and steppe vegetation of Syria, worked out by H. Pabot and quoted by Whyte.[249]

ARABIA

From a phytogeographical point of view, the vegetation of Arabia may be ascribed to the Saharo-Sindian region with a predominance of Mediterranean plants in the north and of tropical species in the south (W. Zeller).[176] Physiognomically the vegetation of Arabia can be divided into:

(*a*) Desert vegetation. Even in the Rub'al-Khali, which apparently receives under 50 mm. of rainfall, some vegetation exists in the form of deep-rooted phanerogams, such as *Calligonum comosum* and *C. torophytes*, as well as *Aristida* spp., *Tribulus* spp. and *Fagonia* spp., whose seeds are able to survive several years' drought.

(*b*) Sub-desert vegetation develops in the areas receiving less than about 100 mm. of rainfall, over most of the Arabian Peninsula, notably on the plains and the sands of Njd and the Hasa and the drier hills of the Hejaz. In central and eastern Arabia much of the sub-desert vegetation is halophytic, especially on the plains where such genera as *Haloxylon*, *Siedlitzia*, *Salsola* and *Suaeda* predominate, while extensive areas of north-eastern Arabia are overgrown by aromatic bushes such as *Artemisia* and *Rhanterium epapposum*. In the foothills, shrubs like *Acacia* sp. and *Lycium* sp. are predominant and grow chiefly in the wadis, where they form associations with tussock grasses such as *Panicum turgidum* and *Cymbopogon schoenanthus*. In the sands of the Nefud and the Dahna, the vegetation is considerably more abundant than in the Rub'al-Khali and also richer floristically, comprising such species as *Calligonum comosum*, *Haloxylon persicum*, *Ephedra alata*, *Artemisia monosperma* and *Rhanterium epapposum*, the last being largely confined to the gaps between the dunes.

(*c*) Steppe vegetation develops in areas receiving additional moisture, through run-off, increased rainfall or perhaps atmospheric humidity, notably in the vicinity of the sea on the coast of the Persian Gulf and the Red Sea. Two major plant associations are the *Rhanterium* steppe vegetation of the Persian Gulf and the *Panicum* steppe vegetation of the Red Sea. In his description of the vegetation of the Red Sea hills and coastal plain south of Jedda, unique in Arabia in being within the tropics and more or less influenced by monsoon rains, D. F. Vesey-Fitzgerald[234] is probably referring to the same type in his tussock-grass savannah on aeolian deposits. The surface of the sandy plain is undulating due to wind action, but no drainage system is developed. Low east–west ridges cross part of the plain, and near the coast small cup dunes occur, which are usually devoid of vegetation and in dynamic condition, with annual herbage in seasons of good rainfall. In hollows and over the more level hummocky areas the climatic climax is tussock-grass savannah in which *Panicum turgidum* and *Lasiurus hirsutus* are dominant, but only small patches remain due to cultivation of *Pennisetum typhoides*. Along the base of the foothills where the substratum is coarser, small scattered trees and shrubs occur. Abandoned cultivated land has a subclimax of *Dipterygium glaucum*.

Annuals, or rather short-lived species, are important members of such communities. The two common grasses are *Stipa capensis* in the north and *Aristida* sp. in the south, and they appear with a variety of other annuals such as *Plantago* sp., *Medicago* sp., *Lotus* sp. and *Fagonia* sp., as well as many others, for a brief period after the rains. In loose sandy habitats, *Stipa* is replaced by *Aristida obtusa* and *Schismus barbatus*. When the annual *Stipa* dries up at the end of the cool season, and if April rains are good, it is replaced by *Aristida plumosa*. Other grasses include *Cymbopogon schoenanthus* and *Chrysopogon aucheri*.[236]

(*d*) Wooded steppe.

(*e*) Xerophilous open woodland.

(*f*) Xerophilous montane vegetation.

Zeller mentions no grasses, and it is difficult to correlate the physiognomic types with the system adopted by Vesey-Fitzgerald.[234, 235, 236] (*g*) Montane woodlands. These are of very limited extent in Arabia, being confined to the small stands of *Juniperus procera* described below, in the highest part of 'Asir and southern Hejaz mountains, but probably not occurring in Yemen. By analogy with the African side of the Red Sea, Zeller suggests that it is possible that below the *Juniperus* belt there once existed a woodland of evergreen *Olea* and *Buxus*. Today, however, it is rare indeed to find even a solitary representative of these genera and the zone supports instead a scrubby growth of *Dodonaea viscosa*, *Tarchonanthus camphoratus*, *Euryops arabicus* and the like. Here and there, especially in the Yemen, there are some sheltered valleys with *Acacia etbaica*.

Vesey-Fitzgerald[234] records the vegetation at different altitudes in this mountain country as follows:

(*a*) Watershed zone: elevations up to 2,000 or 3,000 m. The junipers (*Juniperus procera*) are usually festooned with ichens, while woody herbs (*Euryops arabicus*) and grasses (*Tehmeda triandra*) form a somewhat open understorey.

(*b*) Escarpment and foothills: elevations between 1,500 and 1,000 m. Trees occur on the hillsides at orchard spacing (10–50 m.). *Acacia asak* is abundant, *A. etbaica* frequent, and there are a few deciduous trees, such as *Commiphora schimperi*, *C. myrrha*, *Grewia velutina*, etc. The ground cover is composed of cushion grasses and herbs, and is rich and varied. The common grasses include:

Andropogon distachyus
Aristida coerulescens
A. hirtigluma
A. obtusa
Bromus fasciculatus
Cenchrus ciliaris
Chrysopogon aucheri
Cymbopogon schoenanthus
Danthoniopsis barbata
Eragrostis papposa
Hyparrhenia hirta
Pennisetum setaceum
Tetrapogon villosus

From 1,000 to 500 m., rainfall occurs probably in frequent small showers and run-off is excessive. There is periodic drought. The perennial herbage is somewhat xerophytic; no grass species are mentioned. Below 500 m. *Acacia ehrenbergiana* is the dominant tree in the valleys; tussocks of *Panicum turgidum* occur in sandy-silty places.

Below 300 m., the foothills are much drier and more barren, and the ground herbage is usually scanty.

The remaining physiognomic types of vegetation recognized by Zeller are riparian woodlands, swamps and coastal vegetation.

KUWAIT

The dominant environmental factors influencing the flora and its growth and reproduction are the low seasonal winter, not monsoonal, rainfall (101 mm. average, range from 240 mm. maximum to 26 mm. minimum), high summer temperatures (daily maximum in summer above 44° C, with 20–30 per cent humidity), low winter temperatures (mean for 2–3 months below 15° C with minimum of 0° C or less almost every year), frequent and severe sandstorms, and lack of shade.

M. D. Kernick[117] recognizes three major ecological regions:

(*a*) inland desert of the north and west;

(*b*) low-lying desert plain extending from Al Atraf to the Burgan in the south;

(*c*) the narrow coastal strip, not more than 2–5 km. wide, from Subiya on the north of Kuwait Bay to Ras al Jilai'a on the coast in the south.

The erratic rainfall in the cool season (November to May) and the shortage of moisture in the summer (June to October) restrict the number of plants that can grow. Some plants, mainly annuals, evade drought by completing their growth cycle in a short favourable period and then persist as seeds, bulbs or corms. Others, mainly perennials, develop deep or extensive root systems or reduce loss of water by reduction of exposed surface, production of spines, hairs or waxy coats, and shedding of leaves during dry periods.

The vegetative cover tends to be discontinuous, a mosaic pattern of islands, chiefly composed of simple, mainly single species communities dominated by pure stands of either *Rhanterium epapposum*, *Haloxylon salicornicum*, *Cyperus conglomeratus*, *Panicum turgidum*, *Zygophyllum coccineum* or *Anabasis articulata*, and providing the following breakdown of the vegetation (Kernick's notes on annuals have already been discussed in Chapter III):

(*a*) Coastal saltbush communities

Grasses: *Aeluropus lagopoides*, *Sporobolus spicatus*

(*b*) Coastal sand communities

Chiefly *Panicum turgidum* on top of mounds of wind-blown sand, individual plants often 1·5–3 m. apart, fibrous root system 1·8–2 m. deep, and a spread of 2–2·5 m. in open communities. Associated species are *Elyonurus* (= *Lasiurus*) *hirsutus* and *Rhanterium epapposum*, elsewhere various dicotyledons and *Aristida plumosa*, *Danthonia forsskalii*, *Pennisetum dichotomum*.

(*c*) *Cyperus* steppe

Cyperus conglomeratus, possibly associated with *Rhanterium epapposum* before it was mostly pulled up for fuel, *Aristida plumosa* and *Danthonia forsskalii*.

(*d*) *Rhanterium* steppe

Large area of State composed of pure stands of *Rhanterium epapposum*, with some *Aristida plumosa*.

(*e*) *Haloxylon* steppe

(*f*) *Anabasis* steppe

(*g*) Annual vegetation (see also Chapter III)

Grass species include *Stipa tortilis*, *Sphenopus divaricatus*, *Koeleria phleoides*, *Cutandia memphitica*, *Hordeum glaucum*, *Schismus barbatus*, *Bromus tectorum*.

The shrubby species mentioned above and many others provide much of the grazing for the sheep and goats and/or camels, in addition to the major grasses and *Cyperus conglomeratus*. Reference should be made to Kernick's Report for his record of the dry-matter yields of many of these species, of their seasonal behaviour, of their response to light irrigation with fresh or brackish water. Allowing for a dry-matter intake of 2 kg. per sheep per day, 350,000 hectares of *Rhanterium* range could probably support on a rotational or migratory grazing pattern about 70,000 sheep from November to end February, and 200,000 from end February to end June. For the summer period, 40,000 hectares of *Panicum turgidum* and *Cyperus conglomeratus* could, if protected during winter and spring, support about 30,000 sheep and goats on a rotational basis of 2 hectares per beast from end June to end November. It is stated that *Rhanterium epapposum* reaches a peak protein content of 20–22 per cent during December, January and February—the *digestible* crude protein may, however, not be so high, although the Bedouin report good milk production in their ewes after lambing at this time of the year.

21. Kuwait. *Elyonurus hirsutus* in mixed association with *Panicum turgidum* on deep coastal sands inside an enclosure near Jahara, April 1962. (*Photo M. D. Kernick*)

22. Kuwait. *Cyperus conglomeratus* on windblown sands inside an enclosure near the airport, May 1962. (*Photo M. D. Kernick*)

23. Kuwait. A closely grazed plant of *Cyperus conglomeratus*. (*Photo M. D. Kernick*)

24. North Iran, Elburz mountains. Kandaran Pass, elevation 3000 m. Typical tragacanth vegetation of the mountains of Iran, Afghanistan and Baluchistan, above 3,000–4,000 m. (*Photo H. Pabot*)

25. Central Iran, near Izad-Khast, elevation 2,100 m. Overgrazed, gravelly and silty, trampled steppe. (*Photo H. Pabot*)

26. West Central Iran, 25 km. south-west of Ghom or Qum, elevation 1,200 m. Steppe with skeletal soil, gravelly and silty and/or sandy. (*Photo H. Pabot*)

27. South-west Iran, 27 km. north of Shiraz, elevation 1,700 m. Sub-steppe, mainly stony, silty, calcareous soil, with shrubs uprooted for fuel. (*Photo H. Pabot*)

28. South-west Iran, 32 km. west of Shiraz, elevation 1,950 m. Enclosure protected for two years in sub-steppe vegetation, formerly open oak forest. Soil stony, but fairly deep and calcareous. (*Photo H. Pabot*)

29. South Iran, 120 km. west of Bander Abbas, elevation 200 m. Seriously overgrazed pasture. (*Photo H. Pabot*)

30. South Iran, 25 km. east of Lar, elevation 650 m. Residual grass vegetation in dry wadis. (*Photo H. Pabot*)

IRAN

Iran occupies a unique geobotanical position among the countries of the Near East, acting as a bridge between four major plant geographical regions: Irano-Turanian, Euro-Siberian, Saharo-Arabian and Sudanian (according to the nomenclature adopted by M. Zohary).[263] It is also one of the large speciation centres of the Holarctic desert flora. The climatic diversity is most marked, in both the north to south and west to east directions. The wide variations in topography are as usual responsible for the main differentiation of the vegetation and for the phytogeographic diversity of the flora. There are few countries in Eurasia which have within a latitudinal range of some 11° temperate or even cool temperate forests at their northern limits and tropical (monsoonal) savannah at their southern fringes. Upon the major types of vegetation as governed by temperature, precipitation, topography and soils has been superimposed the influence of man during a span of eight or more millenia. Grazing, browsing, and the collection of plants for industry, fuel, drugs, tragacanth and food since time immemorial has completely changed the original composition of the vegetation and turned it into a series of derived anthropogenic communities.[263]

The flora of Iran contains some 7,000 species, one of the richest in the Near Eastern countries, a remarkable figure for a country of which three-quarters of the surface is poor steppe and desert. M. Zohary finds that, because of the high number of endemics, Iran is one of the centres of speciation for the extraordinarily rich Irano-Turanian flora.

Again according to Zohary, the plant geographical regions are the Euro-Siberian, Irano-Turanian, Mediterranean, Saharo-Arabian and Sudanian. Turanian plants are especially abundant in Khorassan and in central Iran; they may also be found in the vegetation of Arabia, the Negev of Israel, and further west up to the Lybian Desert and the Sahara.

Gramineae: *Stipa lessingiana*, *S. lagascae*, *Aristida pennata*.

The Mediterranean element in Iran is represented by only 0·5 per cent of the total flora. The Sudanian territory is largely Nubo-Sindian, occurring as a strip bordering the Persian Gulf and the Gulf of Oman, extending northwards and eastwards along the wadis which traverse the southern ridges of the Zagros and Makran ranges.

Gramineae: *Cenchrus ciliaris*, *Cymbopogon schoenanthus*, *Dactyloctenium sindicum*, *Dichanthium annulatum*, *Eremopogon foveolatus*, *Pennisetum dichotomum*, *Tricholaena teneriffae*.

This group of species may be compared with those collected by H. Pabot and J. W. B. van der Sprenkel some distance to the west, in the catchment area and plain of Khuzistan, and at other places in southern Iran.

Khuzistan hills and plain: *Aeluropus littoralis*, *A. repens*, *Aristida coerulescens*, *A. plumosa*, *Cenchrus ciliaris*, *Cymbopogon laniger*, *Hyparrhenia hirta*, *Pappophorum persicum*, *Stipa* cf. *barbata*.

Zagros mountains: *Andropogon ischaemum*, *Bromus tomentellus*, *Festuca* cf. *valesiaca*, *Melica* sp., *Oryzopsis* cf. *holciformis*, *Secale montanum*, *Stipa* cf. *barbata*.

South Iran and Isfahan: *Aristida coerulescens*, *Pennisetum orientale*.

A good many species of African origin would presumably have entered Iran at a period when eastern Arabia was connected with Iranian mainland on the one hand, and south-western Arabia with Africa on the other. These connections would have existed until Middle Pliocene ([263]), until which period there could have been a floral interchange between Iran and East Africa via the tropical parts of Arabia.

One of Zohary's vegetation types in Iran is called:

Savannah and Savannah-like Vegetation *Acacietum Flavae* and is characterized by a number of plant communities confined to southern Iran within the Nubo-Sindian territory. The main communities noted are:

(*a*) *Zizyphus spina-christi*/*Calotropis procera*
(*b*) *Z. spina-christi*/*Capparis spinosa* var. *pubescens*
(*c*) *Z. spina-christi*/*Z. nummularia*
(*d*) Acacietum flavae
(*e*) *Amygdalus scoparia*/*Dodonaea viscosa*

The Gramineae in this vegetation type are *Lasiurus hirsutus*, *Cymbopogon schoenanthus* and *Tricholaena teneriffae*.

A. Parsa[164] would recognize nine biotic provinces and ten major types of vegetation, with twenty-seven plant communities.

AFGHANISTAN

According to H. F. Neubauer[157] the country may be divided into three regions which are determined by the Hindu Kush and its prolongation on the one hand, and the watershed between the tributaries of the Indus and of other rivers (with internal drainage) on the other. This watershed has great significance in relation to the monsoonal climate. The monsoon does not, of course, reach the catchment of the Indus tributaries which are much more arid than the regions more to the east, and should be regarded as a transitional zone.

These main regions are (after Linczevski and Prozorovski):

(1) The southern Turkestan region of ephemerals
(2) Central Asian desert region
(3) Afghano/Iranian desert region
(4) Indo/Himalayan forest region.

The altitudinal stratification is fairly clear in all regions (all altitudes are approximate):

(*a*) Lower plains (up to about 500 m.)
(*b*) Higher plains (500–1,000 m.)
(*c*) Lower uplands or mountains (1,000–2,000 m.)
(*d*) Medium uplands or mountains (2,000–2,500 m.)
(*e*) Upper uplands or mountains (2,500–3,000 m.)
(*f*) Subniveous zone (3,000–3,500 m.)
(*g*) Snow zone (3,500–4,000 or 4,200 m.)
(*h*) Altitudes above general vegetation boundary (above 4,300 m.).

The boundaries of the individual zones are not generally very clear and tend to fluctuate, but in eastern Afghanistan, which is influenced by the monsoon and belongs to the Indo/Himalayan forest region, the demarcation is fairly clear. In particular, there is a characteristic dividing line between the various areas of Nuristan and the mountains of Badakhshan, which follows the main crest of the Hindu Kush.

The vegetation of each region, sub-region and altitudinal zone (again, according to Neubauer) exists in two markedly distinct forms, the xeromorphous or xerophilous steppe form, of more or less desertic character which covers most of the country, and the more hygrophilous form, found on the banks of streams, rivers and irrigation channels and in the vicinity of springs.

In general, it may be said that there is a fair degree of correlation between vegetation and climate, although there are examples of areas with the same climate which cannot be equated from the vegetation point of view.

It is interesting to note Neubauer's comments on the vegetation of the *Artemisia* steppe, which is so characteristic of large areas in the Irano/Turanian zone. No doubt the *Artemisia* steppe may be a climax formation in certain regions and under the conditions prevailing there, but it is not always possible to say with any degree of certainty whether it is a climax formation, and in many cases it certainly is not. In Afghanistan it seems to be a sub-type with its own history, and certainly a secondary formation. Centuries of overgrazing by sheep, goats, donkeys and camels have led to the creation of the *Artemisia* steppe in its present form in Afghanistan and to the replacement by it of the original plant communities which may have contained a certain proportion of *Artemisia*. It is probable that the original plant associations were modified by overgrazing in such a way that they now resemble the original *Artemisia* formation which has also been subject to the same modifying factors. Since the *Artemisia* species are grazed only as a last resort, they have increased greatly in number and importance. It has become almost impossible to distinguish between an area of original *Artemisia* association and one which is the result of modifications in a different original plant community. The only way to distinguish between these two types would be to close the whole country to grazing—an experiment which is scarcely feasible.

The Indo/Himalayan forest region corresponds to the eastern part of Afghanistan and may be delimited by a line starting from the point where the Afghan/Pakistani frontier crosses the main Hindu Kush range (in Chitral), along that range beyond the watershed west of the Panjir valley and north of the Ghorband valley, then in a generally southerly direction back to the Pakistan frontier to the east of Wasikha and Ab-i Istada Lake. This line is clearly defined only on the Hindu Kush crest up to the Panjir valley. Farther along, this border is very indistinct, both climatologically and from the point of view of plant geography. The same plant communities continue eastward beyond the Pakistan frontier, especially via Chitral and Kashmir. Or, to be more explicit, that part of the forest region of Afghanistan which is influenced by the monsoon may be considered as the most western section of the forest regions (of all altitudinal levels) located in the southern Himalayan ranges, which taper off on Afghan territory as the monsoonal influence decreases, and gradually merge into the Afghano-Iranian desert region.

The part of Afghanistan which comes under monsoonal influence is the potential forest area of the country, and is largely under tree cover at present. But it can easily be observed how one tree species after another disappears as ecological conditions, particularly general aridity and lack of summer rainfall, gradually intensify to a point where they become intolerable for species after species. The line of disappearance of the forests does not, however, correspond to the theoretical delimitation of this region, i.e. the watershed between the tributaries of the Indus and all other rivers, but runs much further east, so as to exclude the catchments of the Indus tributaries. It roughly follows the line of the Panjir river from the Hindu Kush crest. From its mouth it passes east of Kabul, crosses the mountains east of the Logar valley, passes Gardes approximately across the watershed south of Urgun, and drops into the Gomal valley. The areas west of this line are devoid of forests and, from the point of view of plant geography, should be regarded as belonging to the Afghano-Iranian desert region, or possibly as a transitional zone with much greater affinities to the desert than to the forest (monsoonal) region.

In other words, the border between the dry Irano/Turanian climate and vegetation and the monsoon climate with its characteristic vegetation runs roughly along the border of Pakistan and Afghanistan. Vegetation here follows a catena pattern, in that the slopes of valleys with a southern aspect are monsoonal and those with northern aspect are Irano-Turanian. As is shown by the Japanese work noted below, some of the eastern zones of Afghanistan show a mixture of grass species characteristic of these two contrasting ecoclimates.

In Nuristan, A. Gilli[72, 73, 74, 75] has noted as follows:

In drier spots: *Cynodon dactylon* (in western low hills of Nuristan, and also further west), *Cymbopogon schoenanthus*, *Bromus tectorum* (also further west), *Brachypodium distachyon*, *Carex stenophylla* (also further west).

In moister areas: *Lophochloa phleoides*, *Cynodon dactylon* (spread by irrigation), *Bromus japonicus* (rare), *Hordeum leporinum.*

At higher elevations the flora is alpine, and contains varying proportions of *Kobresia*. At 2,800 m. in Nuristan, *Kobresia capillifolia* represents 50–75 per cent of the total plant cover, together with the rarer *Carex cardiolepis* and *Pennisetum lanatum.*

In deodar forests: *Poa nemoralis*, *P. bulbosa* (the most widespread grass in Afghanistan, even in its var. *vivipara*), and *Phleum graecum.*

The only grass to which Neubauer[157] refers in describing the Indo-Malayan forest range of east Afghanistan is *Desmostachya bipinnata*, which forms extensive grasslands in the Khost basin, particularly near the Matun River between Matun and Dergai.

The relative predominance of Irano/Turanian, Central Asian and monsoonal grass species at high altitudes in the Karakoram and Hindu Kush may be seen from Table 11. In an earlier Flora of Afghanistan, Kitamura[121] showed clearly how that country is a meeting place between the grass species of widely different geographical zones (Table 10)—see also T. Sacco.[194] This latter list contains about 180 species of the Irano/Turanian and Central Asian genera as under:

Aegilops	*Lolium*
Agropyron	*Oryzopsis*
Agrostis	*Phalaris*
Alopecurus	*Phleum*
Bromus	*Poa*
Dactylis	*Puccinellia*
Elymus	*Stipa*
Festuca	*Trisetum*
Hordeum	

TABLE 10

Grass species of the monsoonal vegetation of Afghanistan[121]

Andropogon pertusus	*A. caloptila*
Anthosachne jacquemontii	*A. cyanantha*
A. longe-aristata	*A. plumosa*
Apluda mutica var. *aristata*	*Bothriochloa ischaemum*
Aristida adscensionis	*Brachiaria eruciformis*

B. ramosa
B. reptans
Calamagrostis emodensis var. *breviseta*
C. epigejos
C. munroana
C. pseudophragmites
Chrysopogon aucheri
C. gryllus
C. serrulatus
Cynodon dactylon
Dichanthium annulatum
Digitaria nodosa
Dinebra retroflexa
Echinochloa crusgalli
Eleusine flagellifera
Elyonurus hirsutus
Enneapogon persicus
Eragrostis bipinnata
E. ciliaris
E. megastachya
E. poaeoides
E. tremula
Erianthus fulvus
E. griffithii
E. ravennae
Fingerhuthia affghanica
Heteropogon contortus
Imperata cylindrica
Koeleria phleoides
Panicum antidotale
P. maximum
P. miliaceum
Pennisetum ciliare
P. dichotomum
P. flaccidum
P. lanatum
P. orientale
Phragmites communis
P. karka
Polypogon fugax
P. maritimum
P. monspeliensis
Roegneria nuristanica
R. semicostata
Saccharum halepense
S. spontaneum
Sporobolus commutatus
Themeda anathera
Tragus racemosus

Table 11

Grasses of West Pakistan and Afghanistan, especially Karakoram and Hindukush[122]

Aegilops squarrosa
Baluchistan: Quetta
Agropyron repens
Hindukush: 3,200 m., 3,340 m.
A. schugnanicum
Karakoram: 3,100 m.
A. semicostatum
Hindukush: 2,600 m.
Agrostis canina
Hindukush: 2,600 m., 3,200 m.
A. palustris
Karakoram: 2,500 m.
Hindukush: 2,500 m.
A. semiverticillata
Baluchistan: Quetta
Alopecurus himalaicus
Hindukush: 3,900 m.
A. pratensis
Hindukush: 3,700 m., 4,100 m.
Aristida adscensionis
Karakoram: 2,500 m.
Hindukush: 2,170 m.
Punjab: 900 m.
A. cyanantha
Karakoram: 2,000 m.
Avena fatua
Karakoram: 1,500 m., 2,500 m., 3,200 m.
Hindukush: 2,900 m.
Bothriochloa ischaemum
Karakoram: 1,500 m., 2,500 m.
Hindukush: 2,450 m., 2,500 m., 2,700 m. Karachi

Bromus commutatus
Karakoram: 1,500 m., 3,200 m.
Hindukush: 2,170 m., 2,450 m., 2,500 m., 2,700 m. Karachi
B. inermis var. *villosus*
Karakoram: 3,800 m.
B. tectorum
Hindukush: 3000 m.
Calamagrostis pseudophragmites
Karakoram: 3,500 m.
C. stoliczkai
Karakoram: Urdokas
Chrysopogon aucheri
Baluchistan: Quetta
Cymbopogon schoenanthus
Karakoram: 2,000 m.
Cynodon dactylon
Karakoram: 1,500 m.
Hindukush: 2,250 m., 2,350 m.
Dactylis glomerata
Swat Himalaya: 2,100 m., 2,500 m., 3,050 m.
Punjab: Murree, 2,500 m.
Echinochloa crusgalli
Karakoram: 2,500 m.
Elymus dahuricus
Karakoram: 2,200–2,900 m.
E. dasystachys
Karakoram: 3,300 m.
Hindukush: 2,650 m.
E. giganteus
Karakoram: 3,100 m., 3,300 m.
E. nutans
Karakoram: 3,100 m., 3,800 m.
Hindukush: 2,200–3,650 m.
Elytrigia ferganensis
Karakoram: 3,200 m.
Hindukush: 3,000 m., 3,200 m.
Enneapogon persicus
Hindukush: 2,300 m.
Punjab: Taxila, 900 m.
Eragrostis poaeoides
Karakoram: 2,500 m.
Hindukush: 2,170 m., 2,500 m.
Eremopoa persica var. *songarica*
Hindukush: 2,500–3,100 m.
Eremopyrum buonapartis
Baluchistan: Chaman
Erianthus ravennae
Karakoram: Gilgit Valley
Hindukush: Gupis Valley
Festuca altaica
Karakoram: 3,700 m.
F. ovina
Hindukush: 4,500 m.
Swat Himalaya: between Mahodan and Dishei
F. rubra
Karakoram: 3,100 m.
Swat Himalaya: 2,980 m.
Hordeum leporinum
Punjab: Murree, 2,500 m.
Baluchistan: Quetta
H. turkestanicum
Karakoram: 3,300 m.
Hindukush: 3,000 m.
H. vulgare
Karakoram: Koshimar
Hindukush: 3,100 m.
Koeleria cristata
Hindukush: 3,050–3,900 m.
Leucopoa olgae
Karakoram: 3,600 m.
Lolium perenne
Punjab: Murree, 2,500 m.
L. persicum
Karakoram: 1,500 m.
L. rigidum
Baluchistan: Quetta
Melica jacquemontii
Karakoram: 3000 m.
Oryzopsis brachyclada
Hindukush: 2,500–3,600 m.
O. coerulescens
Baluchistan: Quetta
O. gracilis
Karakoram: 3000 m., 3,200 m.
Hindukush: 2,800 m., 3,200 m.
O. lateralis
Swat Himalaya: 2,060 m., 2,100 m.
Panicum antidotale
Karachi
P. miliaceum
Karakoram: Chapo

Pennisetum flaccidum
Karakoram: 1,500 m., 3,100 m.
Hindukush: 2,200 m., 2,350 m.
Swat Himalaya: 2,200 m.
P. lanatum
Karakoram: Dusso
P. orientale
Karakoram: between Gilgit and Nomal
Baluchistan: Quetta
Phleum alpinum
Swat Himalaya: 2,750–3,200 m.
Ph. himalaicum
Swat Himalaya: 2,100 m.
Phragmites communis
Karakoram: Dusso, Paiju, etc.
Hindukush: 2,350 m., 2,680 m.
Poa alpina
Karakoram: 3,800 m.
Hindukush: 3,900 m.
Swat Himalaya: 3,100–3,650 m.
P. araratica
Hindukush: 3,900 m.
P. bulbosa
Swat Himalaya: 3,200 m.
P. nemoralis
Swat Himalaya: 3,200 m.
P. polycolea
Karakoram: Urdokas
P. pratensis
Karakoram: 1,500 m., 3,300 m.
Hindukush: 3,700 m.
Swat Himalaya: 3,050 m.
P. sterilis
Karakoram: 3,500 m., 3,700 m.
Hindukush: 3,200–2,900 m.
Swat Himalaya: 3,500 m.
P. supina
Karakoram: 1,500 m.
Hindukush: Naz Bar Valley
Polypogon monspeliensis
Karakoram: 2,500 m.
Baluchistan: Chaman
Schismus arabicus
Baluchistan: Chaman
S. barbatus
Karakoram: 1,800 m.
Setaria glauca
Hindukush: Yarkhund
S. italica
Karakoram: Juno
S. viridis
Karakoram: 1,500–3,200 m.
Hindukush: 2,800 m.
Baluchistan: Chaman
Stipa brevifolia
Karakoram: 3,100 m., 3,400 m.
Hindukush: 3,800 m., 4,500 m.
S. caragana
Hindukush: 2,700 m.
S. splendens
Karakoram: 3,600 m.
Hindukush: 2,380 m., 2,600 m.
S. subeffusa
Karakoram: between Minapin and Chalt
Hindukush: 2,350 m., 2,600 m.
S. szowitsiana
Karakoram: 3,400 m.
Baluchistan: Chaman
Taeniatherum crinitum
Baluchistan: Quetta
Tetrapogon villosus
Baluchistan: Quetta
Trisetum spicatum
Hindukush: 3,800 m., 4,460 m.
Triticum aestivum
Karakoram: Koshimar, Askole, Shigar
Hindukush: 2,250–2,800 m.
Vulpia myuros
Hindukush: 2,500 m.

U.S.S.R.

As a contrast to the situation within the transect proper, certain data collected by I. V. Larin[128] from different parts of the Soviet Union are presented in Table 12. As will be seen, this is

TABLE

Changes in the plant cover on virgin lands and on old fallow

Extent of grazing	Flood plains of high Oka meadows (according to N. S. Konyushkov)	Chernozems of Northern Caucasus (according to I. K. Pachovskii)	Common Chernozems of Western Siberia (observations of I. V. Larin)
Absent	—	*Stipa lessingiana*, *Festuca sulcata* with considerable admixture or predominance of motley grass, *Agropyron*, brome grass	*Stipa*, *Festuca sulcata*, *Koeleria*, *Silene*, *Filipendula stepposa*, yellow bedstraw
Light	*Phleum pratense*, *Agrostis alba*, *Geranium pratense*, *Thalictrum minus*, *Pimpinella*	*Stipa*, *Festuca sulcata*, little motley grass and rhizomatic Gramineae	*Stipa*, *Festuca sulcata*, *Koeleria*, *Artemisia frigida*
Moderate	*Festuca rubra*, caraway, *Bromus inermis*, *Alopecurus pratensis*, *Agrostis alba*, *Medicago falcata*	*Festuca sulcata*	*Festuca sulcata*, *Artemisia frigida*
Intensive	*Poa pratensis*, *Taraxacum officinale*, *Achillea millefolium*, *Medicago falcata*, *Carex schreberi*, *Trifolium repens*, *Alopecurus pratensis*	*Poa bulbosa*, much *Carex schreberi*, *Artemisia austriaca*, *Euphorbia seguirana*	*Artemisia frigida*
Excessive	Knotgrass and a few of the above	Knotgrass, *Ceratocarpus arenarius*	Knotgrass
Near cattle farms and wells	—	*Chenopodium tataricum* knotgrass, hedge mustard	Tall coarse plants (hedge mustard, nettles, *Axyris* and others), knotgrass

lands under the influence of different intensities of grazing[128]

Dark chestnut soils of Western Kazakhstan (observations of I. V. Larin)	Ridge-mound sands of Kara Kum (according to O. I. Morozova)	Sandy soils of steppe zone of European RSFSR (according to E. M. Lavrenko)	Clay desert of Central Asia (according to O. I. Morosova)
—	*Arthrophytum, Carex physodes, Salsola richteri, Ephedra, Calligonum, Aristida pennata,* many crusts of mosses, many annuals	Caespitose Gramineae: *Agropyron bekkeri, Koeleria glauca, Stipa ioannis,* admixture of *Artemisia* and of motley grass	*Carex pachystylis, Poa bulbosa, Bromus tectorum, Salsola carinalis, Artemisia herba alba* (in groups)
Stipa lessingiana, Festuca sulcata, Stipa capillata, Koeleria	As above, but fewer annuals and less moss	—	—
Festuca sulcata, Koeleria, Stipa capillata, Artemisia austriaca	*Arthrophytum,* scattered patches of *Carex physodes,* shrubs, *Aristida pennata,* annuals, no moss	Root suckers: *Euphorbia seguirana,* biennials (*Silene,* knapweed and others) and many Gramineae	*Artemisia herba alba* and many of the above plants
Artemisia austriaca, Poa bulbosa, Euphorbia virgata	*Aristida karelini, Salsola richteri,* no sedge, fewer annuals	Rhizomatic: *Agropyron pubiflorum, Calamagrostis epigejos, Elymus giganteus,* more seldom *Carex physodes, Digitaria, Cyrisus* (?) and others	*Peganum harmala* and a few of the above plants
Knotgrass, *Agrypyron* sp., *Ceratocarpus arenarius, Echinopsilon*	Barkhan ring (inland sand formations of a regular semi-lunar form), vegetation very sparse. *Agriophyllum, Aristida karelini, Horaninovia ulicina,* shrubby *Astragalus*	Loose or bare sands; separate thickets of above plants and scattered bushes	—
—	Tyrlo (places where livestock rest), dark patch without plants or sparse small bushes of above plants	—	*Peganum harmala* and infrequent small bushes, *Sophora*

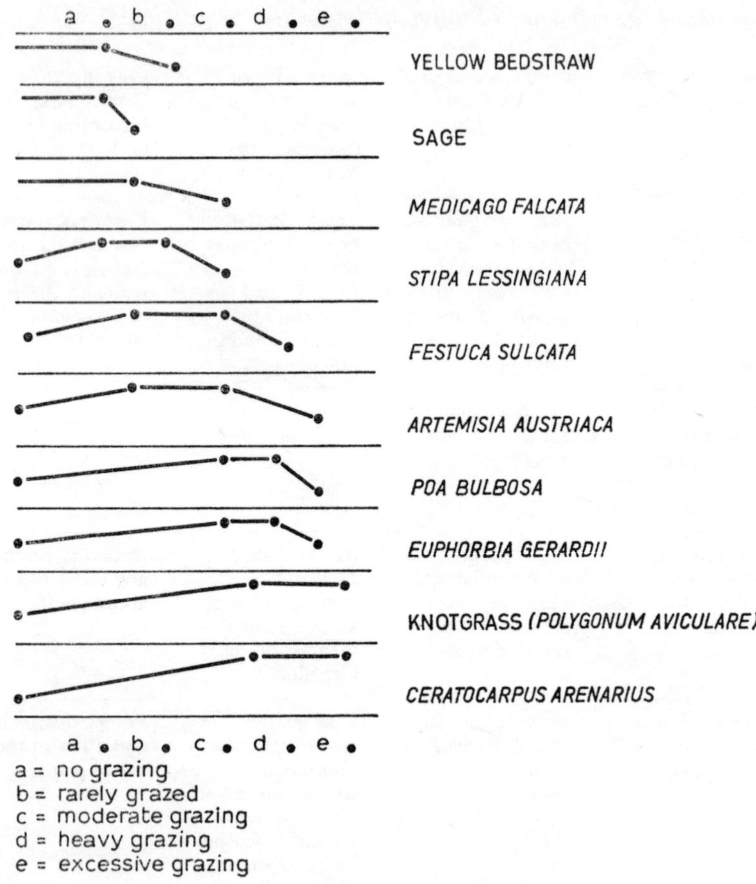

FIGURE 23

Graphical method of showing changes of vegetation under the influence of grazing on virgin land of the southern chernozem of the U.S.S.R., according to I. V. Larin[128]

the world of *Stipa*, *Festuca*, *Agropyron*, *Koeleria* and *Poa*. Larin makes the point with reference to the changes that occur in plant cover under different intensities of grazing that intensification of grazing is not always accompanied by a reduction in the amount consumed by livestock. Frequently, the most productive herbage is that from which the tall species have already been eliminated, a fact which has to be related to the different grazing behaviour and height of harvesting as between cattle and sheep. On virgin stands of *Stipa lessingiana*, the stage of *Festuca sulcata* is more valuable; on coarse grass forest soils, the gramineous stage; on sandy desert soils with shrubby herbaceous vegetation, the stage with *Carex physodes* is the most valuable (a statement with which H. Pabot and others would emphatically disagree). A graphical presentation of the progressive elimination of species under the influence of grazing is shown in Fig. 23.

Chapter VI

SOUTHERN ASIA

INDIA

In his sketch of the flora of British India, Sir Joseph Hooker states that 'in a large sense there is no Indian flora proper', meaning that there is no family of plants peculiar to India, and that even endemic species are rare. In short, India received its flora from surrounding regions. Hooker and Thomson in their 'Introductory Essay to the Flora of India' recognized seven distinct elements, including an African and an Australian element. The fact that Peninsular India has been land since Silurian time is rather an argument against its developing a peculiar flora. Any families or genera it developed have had ample time to spread to adjacent lands, never very far distant.

What is true of India is even more true of Tibet.[243] Peninsular India has been land from the earliest times, but until the end of the Eocene period, Tibet and the Himalaya lay beneath the great Eurasian Sea called the Tethys. Still less then is there a true Tibetan flora, since none could have existed until well into the Tertiary period, and this flora must have been received from adjacent regions.

Again, Tibet was probably completely sterilized during the Pleistocene Ice Age. An ice sheet formed a continuous barrier from the north-west, along the Himalaya to the south-east, completely cutting off the southern peninsula from the interior of Asia, which was a desert. The present Tibetan flora must therefore have entered the country after the retreat of the ice, that is in Pleistocene times, and by existing routes—elements which may be described as eastern Asiatic, Indo-Malayan, European, Mediterranean and northern. In the section on Tibet below is given the list of species compiled by Hemsley over sixty years ago and therefore probably not complete, showing that no less than twenty-nine seed plants are common to Tibet (Sitsang) and the European Alps.

The Tibetan flora can therefore be regarded as a unit, distinct from that of the Central Asian depression described in the section on Tibet. It is possible to suggest that the Tibetan, Himalayan and western Chinese floras really constitute a plant geographical or botanical region—Sino-Himalayan. This suggestion made by Kingdon Ward no longer regards the Tibetan flora as an outlier of the eastern Asiatic region, but attaches it to the Sino-Himalayan, thus rounding this off to the south.

Ecoclimatic Gradients

We have seen how, in West Africa, these gradients run west to east from the Atlantic Coast to the Sudan (Figs. 2 and 3 and Tables 6 and 8). At the author's request, Y. Satyanarayan provided a map showing comparable gradients in India (Fig. 24) together with a Table (13) giving the relevant characteristics of each segment along the gradient. It will be seen that rainfall increases from west to east in Rajasthan and adjacent parts of western India, and from east to west from the low rainfall zones of the Deccan in Peninsular India to the coast along the Western Ghats.

Ecological Status of Grass Cover

Grassland in a broad, non-physiognomic sense may be said to be that land on which gramineous species represent the dominant if not the exclusive vegetation. Not all that is in India called forest or which comes under the Forest Administration is really forest; many such areas are grasslands. When grasses no longer predominate and trees get the upper hand, we have woodland or forest. There are hardly any artificial grasslands in India, certainly none in the forests, but the existing grasslands are not wholly natural either. In an indirect way, they owe their existence to human activities. The consensus of opinion among ecologists is that, but for the lopping, shifting cultivation, burning and grazing practised through the ages in the forests, there would probably be no grasslands as such in India.[252, 271]

Ecologists generally conclude that, as a climax formation, grassland is intermediate in status between forest or woodland on the one hand and desert on the other. In India, however, while woodland and desert occur as climax formations, climax grassland, in the true sense, equivalent to steppe, range, pampa

TABLE

Ecoclimatic gradients in India

Zones	Approximate rainfall range, cm.	Dry season, months	Vegetation	Forest formation	Chief tree species
Arid	30 and below	10–11	Desert	Scrub	a. *Acacia jacquemontii* *Calligonum polygonoides* *Calotropis procera*
Semi-arid	30–60	7–9	Thorn forests	Open scrub woodland	(i) *Prosopis spicigera* *Salvadora oleoides* *Acacia leucophloea* *Acacia arabica* *Capparis decidua* *Balanites aegyptiaca* *Zizyphus nummularia*
					(ii) *Acacia senegal* *Anogeissus pendula* *Cordia rothii*
					(iii) *Acacia leucophloea* *Acacia arabica* *Acacia latronum* *Capparis* and *Zizyphus* spp.
					(iv) *Acacia planifrons* *Acacia latronum* *Albizzia amara*
				Savannah woodland	(v) *Pterocarpus marsupium* *Anogeissus latifolia* *Acacia latronum* *Albizzia amara*
Dry sub-humid	60–120	6–8	Woodlands, savannahs	Open dry deciduous types (i) Dry teak	*Tectona grandis* *Adina cordifolia* *Emblicata officinalis*
				(ii) Dry *sal*	*Shorea robusta* *Pterocarpus marsupium* *Terminalia tomentosa* *Anogeissus latifolia*

13

prepared by Y. Satyanarayan)[253]

;rass cover	Main types of land-use	Main crops	Livestock	State
asiurus sindicus–Cenchrus pp.–Panicum spp. *P. turgidum* and *P. anti-'otale*)	Nomadic and semi-nomadic grazing, dry farming	Millets, sorghum barley	Camels, sheep, goats, cattle and donkeys	Rajasthan
'enchrus–Dichanthium unnulatum, Eremopogon oveolatus, Eleusine com-ressa–Dactyloctenium indicum, Aristida	Nomadic and semi-nomadic grazing, intensive dry farming, utilization of trees and shrubs for grazing, also cultivation with well irrigation, dairy industry (northern India), fallows of 3 to 7 years' duration	Lesser millets, (Eleusine) millets, sorghum, barley, cotton, oil seeds, some wheat	Camels, sheep, goats, cattle, some buffaloes, donkeys (note: goats are for mohair and mutton)	Rajasthan, Punjab, Gujarat, Maharashtra and Madhya Pradesh
'ymbopogon jwarancusa-'ehima nervosum	Intensive dry farming, multiple cropping, free grazing, utilization of forests and minor products for fuel and timber, fallows of 2 to 3 years (southern India)	As above, besides groundnuts, cotton, irrigated rice, sugar cane, chillies, wheat and pulses	Sheep, cattle, goats, some buffaloes	Maharashtra, Mysore, Andhra Pradesh, Madras
)ichanthium annulatum–'hemeda quadrivalvis, 'seudanthistiria heteroclita				
Ieteropogon contortus, ;ristida redacta, Aristida ıniculata				
'ymbopogon coloratus–'hemeda triandra				
'ymbopogon coloratus Ieteropogon contortus–'hemeda triandra, Iseilema ıxum, Chrysopogon mon-ınus, Aristida spp.	Dry and irrigation farming, seasonal fallows, horticultural crops, utilization of forest products, forest grazing for local consumption and export	Millets, sorghum maize, cotton tobacco, oil seeds, wheat, pulses, mangoes, citrus	Cattle, buffaloes, goats, some sheep, horses, donkeys	Punjab, Madhya Pradesh, Gujarat, Maharashtra, Andhra Pradesh, Madras, Uttar Pradesh
'hrysopogon montanus 'othriochloa intermedia 'ulaliopsis binata				

Table 13—continued

Zones	Approximate rainfall range, cm.	Dry season, months	Vegetation	Forest formation	Chief tree species
Moist sub-humid	120–160	5–8	Open woodlands with eastern areas covered by *sal* and western areas by teak and *Terminalia*	Seasonal forests with mixed evergreen and deciduous species (i) Moist teak	*Tectona grandis* *Terminalia crenulata* *Terminalia paniculata* *Terminalia bellerica* *Dalbergia latifolia* *Lagerstroemia lanceolata* *Grewia tiliaefolia*
				(ii) Moist *sal*	*Shorea robusta* *Lagerstroemia parviflora* *Dillenia pentagyna* *Pterocarpus marsupium* *Syzygium cumini* *Schima wallichii* *Terminalia tomentosa*
Per-humid	160–200	2–6	Moist forest at low and medium altitudes	Evergreen *seasonal* forest (western form)	*Actinodaphne angustifolia* *Olea dioica* *Pouteria tomentosa*
				(eastern form)	*Cinnamomum cecidodaph* *Amoora wallichii* *Mesua ferrea* *Altingia excelsa* *Manlietia insignis* *Artocarpus chaplasha* *Eugenia* spp. *Tetrameles nudiflora* *Stereospermum chelonioic*
Wet	300–400	2–4	Wet forests at medium altitudes; optimum type	Closed rain forest (western form)	*Poeciloneuron indicum* *Palaquium ellipticum* *Mesua ferrea* *Dipterocarpus indicus* *Hopea parviflora* *Cullenia excelsa* *Vateria indica*
	200–300	2–3	Wet forests at medium altitudes; optimum type	(eastern form)	*Calophyllum elatum* *Dipterocarpus pilosus* *Artocarpus chaplasha* *Artocarpus heterophyllus* *Mesua ferrea* *Cinnamomum cecidodaph* *Altingia excelsa*

ass cover	Main types of land-use	Main crops	Livestock	State
udanthistiria heteroclita *emeda quadrivalvis* *chanthium annulatum* (in ins) *undinella pumila* *ima nervosum* (on the s)	Some shifting cultivation, forest utilization on scientific basis, arable cultivation with seasonal fallows, forest grazing, horticultural crops	Rice, jute, sugar cane, pulses, tobacco, mangoes bananas	Cattle, buffaloes, goats, some sheep	Maharashtra, Mysore, Orissa, Bihar, U.P. and Bengal
perata cylindrica, Narenga phyrocoma, Saccharum ntaneum, Themeda ndinacea, Microstegium atum				
undinella–Cymbopogon	Shifting cultivation, intensive cultivation, short fallows, forest utilization	Jute, rice, pulses, wheat, sugar cane cinchona, rubber cashewnut (in W. Ghats), coffee (Coorg)	Cattle, buffaloes, goats	Maharashtra, Mysore, Kerala, Assam and Bengal
perata–Saccharum–emeda *ragmites–ccharum*				
grass in the undisturbed est	Shifting cultivation, spice gardens, forest utilization for timber and forest products, plantation crops, horticultural crops	Hill rice, tapioca, cardamom, pepper, tea, rubber, bananas, jack fruits, coconuts on the coast	Cattle, buffaloes, goats, pigs, horses	Mysore, Kerala, Assam
grass in the undisturbed est				

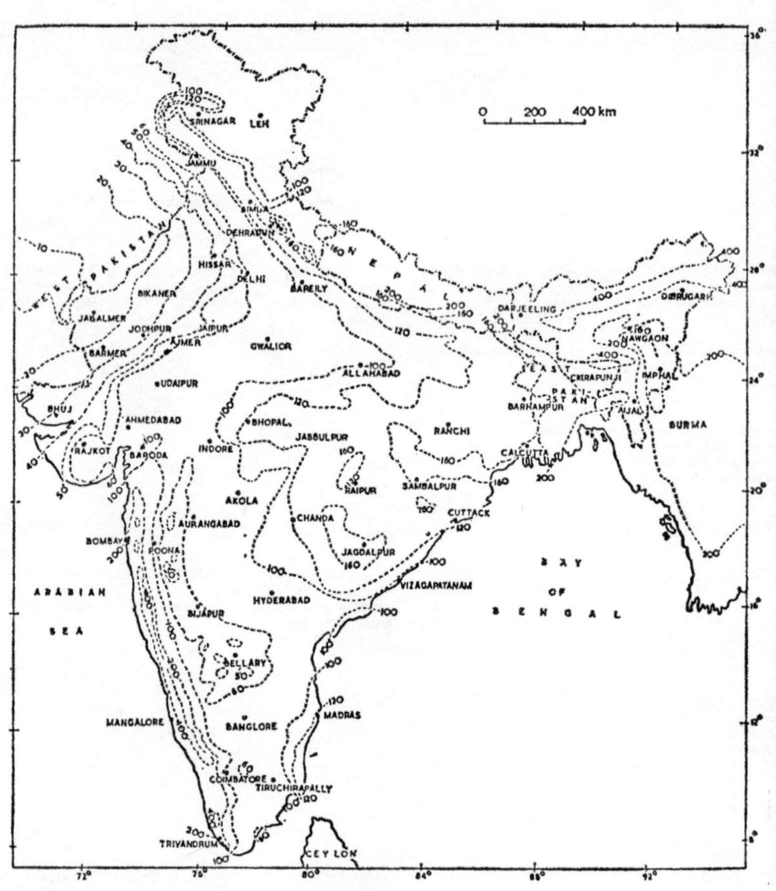

FIGURE 24

Rainfall map of India; isohyets are in cm. (*From Map 6 of 'National Atlas of India'*, preliminary edition)[253]

or savannah, appears to be absent. Doubts are now expressed by grassland ecologists in other parts of the world regarding the authenticity of these long-accepted grassland climaxes. N. L. Bor[18] considers that the climate in India is either a forest climate or a desert climate. H. G. Champion[30] also doubts the existence in India of any examples of climax grasslands, although grassland is common enough as an interim stage in the succession (seral stage) and sometimes may even be relatively very stable (pre-climax) under the influence of repeated fire and grazing. The typical savannah type of grassland which is believed to occur as a climatic climax in other parts of the tropics and sub-tropics is apparently absent as such in India. The closed deciduous forest grades into thorn forest without an intermediate open park-like stage and there is no steppe or range type of grassland climax between the woodland and the desert. Even the alpine grasslands have been subjected to such intensive summer grazing for centuries that we cannot venture to suggest what species might be represented in the truly climax formations there.

Locally in India, grassland may be found as an edaphic climax on wet soils and in frost pockets in the sal belt and in the hills in other parts of the country, frost and excessive moisture inducing a grassy climax, but these circumstances do not make these areas a climatic climax. Certain lower alluvial deposits carry extensive grasslands, some of which are believed to be a stage in the primary succession, particularly those in the wetter and drier sites which trees cannot colonize. *Saccharum spontaneum* dominates the beginning of primary succession on sandy alluvial deposits. More generally, however, the grasslands on the lower alluvial deposits are a sequel to felling, burning and grazing; those on the higher alluvium are also biotic in origin, being the result of human settlement and destruction of forests. The regeneration of forest is prevented by fire, grazing and cutting, and by general deterioration of the soil. Extensive areas in the sal belt which have suffered this treatment in the past now carry high savannah grasses, e.g. on the higher phantas of Kheri and the chaors of Haldwani in Uttar Pradesh.

There are extensive grass downs on the Nilgiri, Palni and Anamalai plateaux in southern India where grazing and burning have been regularly practised. Discussing the ecology of the Nilgiri grasslands, C. R. Ranganathan[182] argues that even more intensive grazing and burning elsewhere have not produced similar grasslands, that these Nilgiri grasslands owe their origin to frost as a major climatic factor, and that they should, therefore, be regarded as a climatic climax in common with the local sholas (temperate wet evergreen type of forests). On the other hand, N. L. Bor[18] contends that once the existence of grazing and burning is admitted, the term 'climatic climax' cannot be applied to the community under any system of terminology;

these grasslands can, therefore, be considered only as a biotic climax. A recent statement of succession in these grasslands is given below (Fig. 29).

With the existing evidence, it seems correct to conclude that grasslands do not occur as a climatic climax in India; that they generally represent secondary seres and rarely also priseres; and that, wherever they are relatively more stable, they constitute a biotic climax or, in Clementsian terminology, a disclimax more frequently of the nature of a sere-climax than of a sub-climax maintained under the influence of grazing, cutting and burning in varying intensities and combinations, together with shifting cultivation.

Grasslands of India

Until fifteen years ago, Indian and British botanists had done invaluable work on the recognition, naming and classification of the great number of grass species in the Indian flora, and N. L. Bor[19] has produced a major work on the subject. Comparatively little attention had, however, been given to the grass communities, to their botanical composition and to the nature of the succession that occurred under the influence of the various biotic factors. Botanical and ecological studies are of great importance in the correct understanding of India's grasslands. What may be called ecological management of grassland, the manipulation of natural succession, is the chief if not the only method which can be adopted for the improvement and maintenance of grazing grounds, village pastures, forest grazings and similar uncultivated and uncultivable areas. It was to obtain the essential basic information that the Indian Council of Agricultural Research decided in 1953 to undertake a rapid reconnaissance survey of the grasslands of India.[254]

Two members of the survey team received training under FAO fellowships in the United States of America and Great Britain, and others were trained at the Forest Research Institute, Dehra Dun. The several items of the survey and the techniques adopted on selected one-acre plots were as follows:

(*a*) FLORISTIC COMPOSITION

(i) listing and classification of species occurring in the community,

(ii) estimates of percentage composition by the pace transect method.

(*b*) DENSITY OR PLANT COVER
judged by the square foot density method[213] from a circular sample enclosing an area of 100 sq. ft.

(*c*) FORAGE PRODUCTION
estimated by cutting and weighing the individual species from ten 100 sq. ft. density circles.

(*d*) PLANT VIGOUR
assessed by the study of plant height, leaf length, length of seedstalk, number of tillers, basal area and reproduction of the species.

(*e*) PLANT SUCCESSION
data obtained by inference and by a comparison of floristic composition on undisturbed and overgrazed areas.

(*f*) FOREST TYPES
listing of associated tree species and classification of the forest types on the basis of H. G. Champion.[30]

(*g*) SOIL TYPES
data collected according to the method adopted in soil surveys, for colour, texture, water-holding capacity, pH, total soluble salts, and soluble phosphate and potash by rapid testing methods.

P. M. Dabadghao and his colleagues have recognized a number of grass cover types. Particulars of these, extracted from Chapter 5 of *The Grass Cover of India*,[254] will now be given.

Types of Grass Cover

An examination of the distribution pattern of the various grass covers shows that their occurrence is primarily governed by climatic factors, and chiefly by latitudinal influence. Thus the tropical *Sehima/Dichanthium* cover is quite distinct from the sub-tropical *Dichanthium/Cenchrus/Lasiurus*, *Phragmites/Saccharum/Imperata* and *Themeda/Arundinella* covers. The latter are again quite different from the temperate alpine cover.

The second factor, in order of importance, appears to be topography, more particularly altitude. This is shown clearly by an examination of the distribution of the sub-tropical grass covers, two of which, namely, *Phragmites/Saccharum/Imperata* and *Dichanthium/Cenchrus/Lasiurus* occur in the plains, while *Themeda/Arundinella* is restricted to the northern hills.

In the distribution of the grass covers in the northern montane region, the altitude appears to accentuate the latitudinal influence. Thus the *Themeda/Arundinella* type is restricted to a maximum altitude of about 2,200 m. within the same latitude; beyond that altitude only species of the temperate alpine cover exist. As indicated in the discussion of these two last covers, exposure or aspect does not appear to play a significant role. However, it is not impossible that it may have a localized influence on composition under similar use factors.

After climate and topography the soil factor, and more particularly the soil moisture relationship, seems to be important in determining the occurrence of grass covers. Thus in the northern plains, increase in soil moisture, which makes the habitat almost hydrophytic, determines the distribution of the *Phragmites/Saccharum/Imperata* type. On the other hand, the gradual deterioration in soil moisture from optimum through critical towards precarious conditions governs the occurrence of the *Dichanthium/Cenchrus/Lasiurus* type. In this cover, wherever soil moisture rises in excess of requirements, species of the *Phragmites/Saccharum/Imperata* type invariably occupy the site.

For each of the following grass cover types, particulars are given in *The Grass Cover of India* of distribution, environment, grass vegetation (perennials, annuals), other herbaceous vegetation, forest types,[31] successional trends, management. Only some of these aspects are given below.

Sehima/Dichanthium Type

This cover type spreads over the whole of Peninsular India, including the Central Indian Plateau, the Chhota Nagpur Plateau and the Aravalli Ranges, comprising the States of Gujarat, Maharashtra, Madhya Pradesh, Orissa, Andhra Pradesh, Mysore, Madras, Kerala, and also south-west Bengal, southern Bihar and the southern hilly portions of Uttar Pradesh and Rajasthan, with a potential coverage of approximately 1,740,000

31. West Pakistan. *Chrysopogon* type far from water near Loralai, in former Baluchistan. (*Photo A. Johnston*)

32. West Pakistan. *Elyonurus hirsutus*, with scattered plants of *Eleusine flagellifera*, on a protected site near Dera Ghazi Khan (*Dichanthium/Cenchrus/Elyonurus* type). (*Photo A. Johnston*)

33. West Pakistan. Deteriorated rangeland near Rawalpindi. (*Photo A. Johnston*)

34. India. *Cenchrus ciliaris* and *C. setigerus*, coming in due to protection and soil accumulation provided by bushes of *Zizyphus nummularia* in a grassland where *Aristida* predominates. (*Photo Mahendra Prakash*)

35. India. *Cenchrus ciliaris* and *C. setigerus* in a sown pasture in Jodhpur District, sandy loam soils near Sardarsamand. Scattered trees of *Prosopis spicigera*. (*Photo Mahendra Prakash*)

sq. km. The cover is also found in the coastal region. The cover type is associated with a variety of recognized soil types, namely 'mixed red and black', 'medium deep black', 'red and yellow', 'red gravelly', 'red loam' and 'laterite'.

The following 24 species are the most characteristic perennials of this grass cover. Each of them is the dominant species in one of the 24 grass communities representing different stages of development within the major *Sehima/Dichanthium* type.

Aristida setacea
Arundinella mesophylla
Bothriochloa pertusa
Chloris dolichostachya
Chrysopogon fulvus
Ch. orientalis
Cymbopogon coloratus
C. gidarba
C. jwarancusa
C. martinii
C. travancorensis
Cynodon dactylon
Dichanthium annulatum
D. caricosum
D. nodosum
Eragrostis coarctata
Eremopogon foveolatus
Eulalia trispicata
Heteropogon contortus
H. triticeus
Ischaemum indicum
Iseilema laxum
Sehima nervosum
Themeda triandra

There are also many associated perennial species contributing more or less than 10 per cent in the frequency percentage, many annual grasses, and 129 herbaceous species.

Dichanthium/Cenchrus/Lasiurus Type

This type is associated with sub-tropical arid and semi-arid regions comprising the northern portion of Gujarat, the whole of Rajasthan, excluding the Aravalli ranges in the south, western Uttar Pradesh, Delhi State and Haryana, with a potential coverage of more than 436,000 sq. km. On the southern boundary of the type a mingling of the grass species of the *Sehima/Dichanthium* and the *Dichanthium/Cenchrus/Lasiurus* covers takes place; this could therefore be considered as a transitional zone between the two types.

The associated soil types are: 'undifferentiated alluvial', 'grey and brown soils of the Indus and Jumna basin impregnated with salts', and the 'cover grey and brown soils of the desert'. On the basis of the soil data collected during the Survey, the soil characteristics associated with the grass cover are as follows:

The soil colour ranges from light yellowish brown through grey-brown to almost grey, as in soils with impeded drainage. The soils are generally well provided with phosphate, while the potash content is low to medium.

FIGURE 25

India. Succession in the *Sehima/Dichanthium* cover

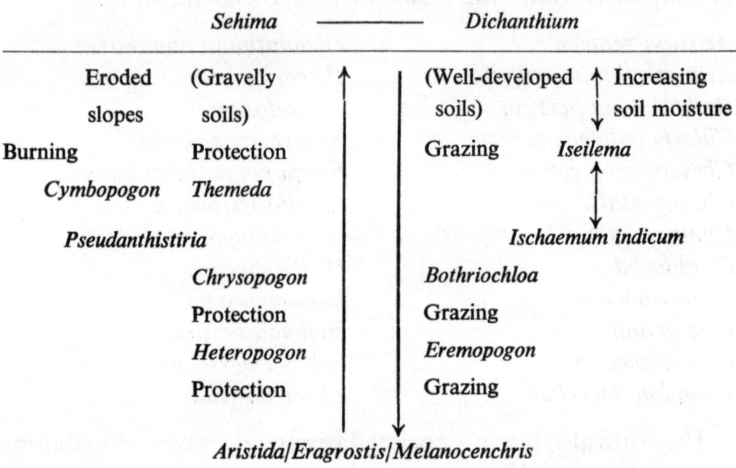

The principal perennial species characterizing this grass cover are the following (as stated under the *Sehima/Dichanthium* type, each of these species is the dominant of the grass communities at different stages of development within the major *Dichanthium/Cenchrus/Lasiurus* cover).

Cenchrus ciliaris
C. setigerus
Cymbopogon jwarancusa
Cynodon dactylon
Dactyloctenium sindicum
Desmostachya bipinnata
Dichanthium annulatum
Eleusine compressa
Eremopogon foveolatus
Lasiurus sindicus
Sporobolus marginatus

Phragmites/Saccharum/Imperata Type

This cover type occurs throughout the Gangetic Plain, the Brahmaputra valley and extends westwards into the plains of the Punjab; the area comprises approximately 2,800,000 sq. km. in the States of Manipur, Assam, Tripura, West Bengal, Bihar, Uttar Pradesh, Delhi and Haryana.

FIGURE 26

India. Succession in the *Dichanthium/Cenchrus/Lasiurus* cover

Dichanthium/Cenchrus/Lasiurus

↑	↓
Protection	Grazing
	Cenchrus/Lasiurus
Protection	Grazing
	Cynodon/Eleusine
Sporobolus marginatus	Grazing
Protection	*Aristida*
Chloris (Compact soil)	*Cenchrus biflorus* (Loose soil)

Recognized soil groups associated with this type are 'undifferentiated alluvial soil', 'grey and brown soils of the Indus and Jumna Basin impregnated with salts', and 'Gangetic alluvium (calcareous)'.

In the semi-arid region, the type appears to be represented by only 6 perennial species, in contrast with the wide variety of species found in the high-rainfall region. These six species are:

Desmostachya bipinnata	*Saccharum arundinaceum*
Imperata cylindrica	*S. bengalense*
Phragmites karka	*S. spontaneum*

The first-named appears to be the indicator species of the type in the semi-arid tract. The soil reaction in this case is generally neutral to slightly alkaline, while it is strongly acidic in the high-rainfall region.

The principal species representing this grass cover throughout its area of distribution are:

Bothriochloa intermedia
B. odorata
Chrysopogon aciculatus
Cynodon dactylon
Desmostachya bipinnata
Hymenachne pseudointerrupta
Imperata cylindrica
Ischaemum timorense
Narenga porphyrocoma
Neyraudia reynaudiana
Panicum notatum
Paspalum conjugatum
Phragmites karka
Saccharum arundinaceum
S. bengalense
S. spontaneum
Sclerostachya fusca
Sporobolus indicus
Vetiveria zizanioides

As in the two preceding cover types, each of the principal species represents a grass community. In other words, the cover is represented by 19 grass communities in various stages of development.

FIGURE 27

India. Succession in the *Phragmites/Saccharum/Imperata* cover

Phragmites/Saccharum/Imperata (Savannah)
↑ ↓ ↓
Protection — Burning and cutting — Grazing
↑ ↓ ↓
Saccharum/Imperata/Sclerostachya — *Vetiveria*
↑ ↓
Protection — Burning and grazing
↑ ↓
Desmostachya/Imperata (depauperate)
↑ ↓
Sporobolus/Paspalum/Chrysopogon

Themeda/Arundinella Type

This cover type occurs in the entire northern and north-western montane tract, on an area of approximately 230,400 sq. km. in the States of Manipur, Assam, West Bengal, Uttar Pradesh, Punjab, Himachal Pradesh and Jammu and Kashmir. Coming from the plains, a change in the direction of the type takes place approximately at an altitude of 350 m. above sea level. The upper limit for this cover appears to be around 2,100 m., above which a decided temperate influence is evinced by the preponderance of the temperate grass species. In general, the belt between 1,800 and 2,100 m. is considered as a transitional zone.

The rainfall in the tract ranges from about 1,000 mm. in the western region to over 2,000 mm. in the east, and as much as about 12,500 mm. at Cherrapunji in Assam. Mean June temperature is of the order of 27°–32° C, while mean January temperature has a range of 13°–17° C. Snowfalls are quite a common feature in winter in most of the areas above 1,200 m. Depending upon elevation, the snow remains on the ground from a few days to about 4 months, and disappears by March or April. The type is associated with 'undifferentiated forest and hill soils', and also with 'undifferentiated sub-montane regional soils'.

The principal perennial species characterizing the type are:

Arundinella bengalensis
A. nepalensis
Bothriochloa intermedia
B. pertusa
Chrysopogon fulvus
Ch. gryllus
Cymbopogon jwarancusa
C. olivieri
Cymbopogon stracheyi
Cynodon dactylon
Dimeria fuscescens
Eragrostiella leioptera
Eulaliopsis binata
Heteropogon contortus
Ischaemum barbatum
Themeda anathera

Each of the above species is the dominant of the grass communities at different stages of development within the major grass cover type.

FIGURE 28

India. Succession in *Themeda/Arundinella* cover

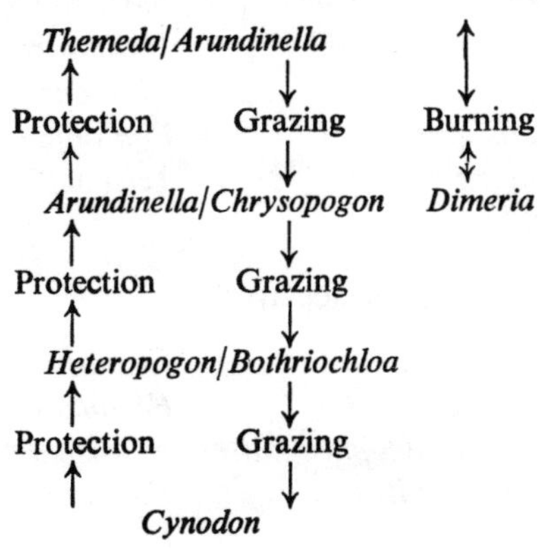

Temperate Alpine Type

Survey data are insufficient for definite conclusions to be made regarding this type, since the team recorded only 25 sites in Tehri-Garwhal and North Garwhal, 4 sites in Jammu and Kashmir, 2 in Himachal Pradesh and 4 in West Bengal. The type occurs at elevations above those reached by the *Themeda/Arundinella* type, 2,100 m. and above in the west, and 1,500 m. and above in the east of the whole Himalayan region. Rainfall varies from 375 mm. in the Chini area in Himachal Pradesh, to 3,750 mm. in the Darjeeling district of West Bengal. Snow may remain for 4–6 months or longer, depending upon elevation. This cover type is associated with 'undifferentiated forest and hill soils' and 'undifferentiated submontane regional soils'.

The grass vegetation of this primarily non-monsoonal cover is composed of the following species:

Agropyron canaliculatum
Agrostis canina
A. filipes
A. munroana
A. myriantha
Andropogon tristis
Calamagrostis epigejos
Chrysopogon gryllus
Dactylis glomerata
Danthonia jacquemontii
Koeleria cristata
Phleum alpinum
Poa pratensis
Stipa concinna

Associated perennial species with a contribution of more than 10 per cent in the composition are:

Agrostis pilosula
Brachypodium sylvaticum
Bromus ramosus
Calamagrostis emodensis
Eragrostis nigra
Festuca lucida
Festuca valesiaca
Helictotrichon asperum
Muhlenbergia sp.
Poa alpina
Trisetum sp.

Other perennial species are:

Agropyron semicostatum
Agrostis micrantha
Calamagrostis pseudophragmites
Deschampsia caespitosa
Deyeuxia scabrescens
Erianthus longisetosus
Festuca gigantea
F. kashmiriana
F. ovina
F. rubra

Glyceria tonglensis	*Oryzopsis aequiglumis*
Helictotrichon pratense	*Poa angustifolia*
H. virescens	*P. pagophila*
Milium effusum	*Stipa sibirica*
Muhlenbergia duthieana	*Trisetum micans*
M. huegelii	*T. spicatum*

The annual grass species associated with the type are:

Oryzopsis lateralis	*Poa stewartiana*
Poa annua	*Polypogon fugax*
P. jaunsarensis	

The forest types associated with these temperate alpine grass communities are:

(1) Himalayan chir pine forest
(2) Moist deodar forest (*Cedrus*)
(3) Himalayan sub-tropical pine forest
(4) Birch-rhododendron scrub forest
(5) Deciduous alpine scrub
(6) Alder forest
(7) East Himalayan sub-tropical wet hill forest
(8) East Himalayan mixed coniferous forest.

Nilgiri high-altitude type

At elevations around 2,300 m. in the Nilgiris in southern India (latitude 12° N), *Andropogon polyptychus* (*Dichanthium polyptychum*) is the 'grassland climax' with *Chrysopogon zeylanicus* a subclimax stabilized by biotic factors such as fire and grazing

TABLE 14

Percentage composition of typical 'Chrysopogon zeylanicus' grasslands near Mukurthi Reservoir, Nilgiris, not subject to overgrazing

	Per cent
Chrysopogon zeylanicus	44
Ischaemum indicum	18
Themeda triandra	6
Andropogon lividus	4
Cymbopogon polyneuros	2
Arundinella setosa	2
Weeds	10
Blanks	9

and the resultant severe erosion.[77], [78] Studies of progression and regression show that these grasslands are a montane phase of the broad *Sehima/Dichanthium* type of Peninsular India, recognized by the Grassland Survey.[254]

TABLE 15

Percentage composition of grasslands near Avalanche, Nilgiris, not grazed

	Per cent
Andropogon polyptychus var. *deccanensis*	61
Eulalia phaeothrix	15
Themeda triandra	10
Andropogon lividus	7
Tripogon bromoides	Trace
Ischaemum indicum	Trace
Weeds	3
Blanks	4

FIGURE 29

India. Succession in Nilgiri grasslands

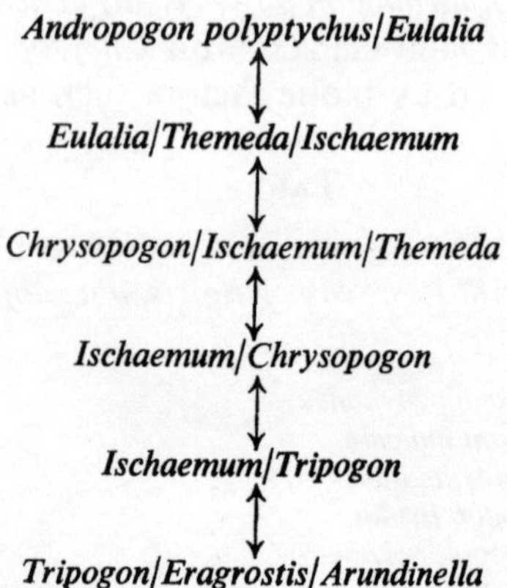

Succession in vegetation of Western Rajasthan

The present natural vegetation of Western Rajasthan, which is of great comparative interest in the present context, is a sparse low cover of thorny woodland formation with distinct variants in different topographical sites. Distinctive communities are found on different habitats. The original natural vegetation has been disturbed almost everywhere apart from a few relict areas. Almost all communities are of a disclimax or bioclimax nature. Y. Satyanarayan has worked out the course of succession in three different land forms (Figs. 30–32).

FIGURE 30

India. Succession in vegetation of Western Rajasthan

(deep fine sands) SAND PLAINS		(deep fine sands) SAND DUNES
	Prosopis spicigera/Salvadora oleoides	
Gymnosporia spinosa *Balanites aegyptiaca*	↑ ↑	*Prosopis spicigera* *Tecomella undulata*
		Acacia jacquemontii *Gymnosporia spinosa* *Balanites aegyptiaca*
Cymbopogon spp. *Leptadenia pyrotechnica* *Crotalaria burhia* *Lasiurus sindicus* *Panicum antidotale*		*Calligonum polygonoides* *Panicum turgidum* *Leptadenia pyrotechnica*
Cenchrus catharticus *Dactyloctenium sindicum*		*Lycium barbatum* *Crotalaria burhia* *Aerua persica*
Mollugo nudicaulis *Indigofera argentea*	↓ ↓	*Indigofera argentea* *Cyperus arenarius*

FIGURE 31
India. Succession in vegetation of Western Rajasthan

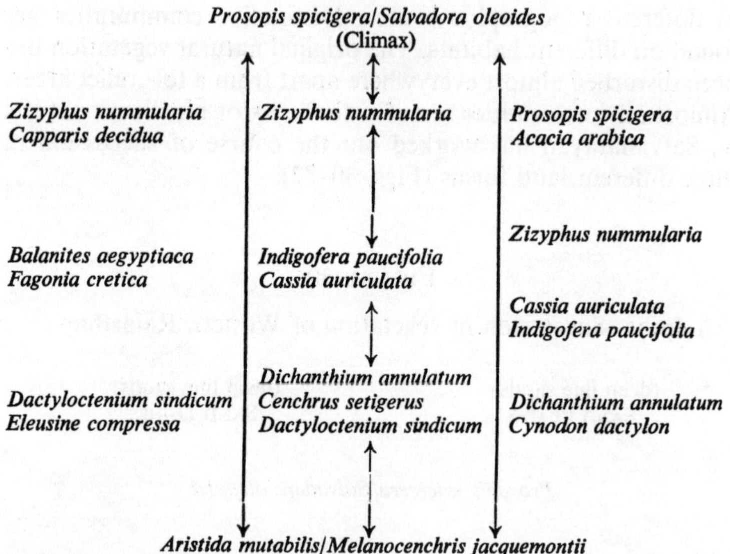

WEST PAKISTAN

Since no information has been found relating to the grass species and communities of East Pakistan, this account has necessarily to be restricted to West Pakistan. It is probable that the species of the east will be similar to those reported by the Grassland Survey of India,[254] for the plains and for the hills and mountains around Darjeeling and Kalimpong, in West Bengal, and for a number of sites in Assam. The grass flora of the mountains of Manipur and Tripura on the eastern border of East Pakistan may possibly show similarities with that of the higher altitudes in Burma (p. 202).

The vegetation of West Pakistan presents a picture of great complexity, comprising vegetation types which may be loosely called desertic, Mediterranean, tropical and temperate (A. John-

FIGURE 32

India. Succession in vegetation of Western Rajasthan

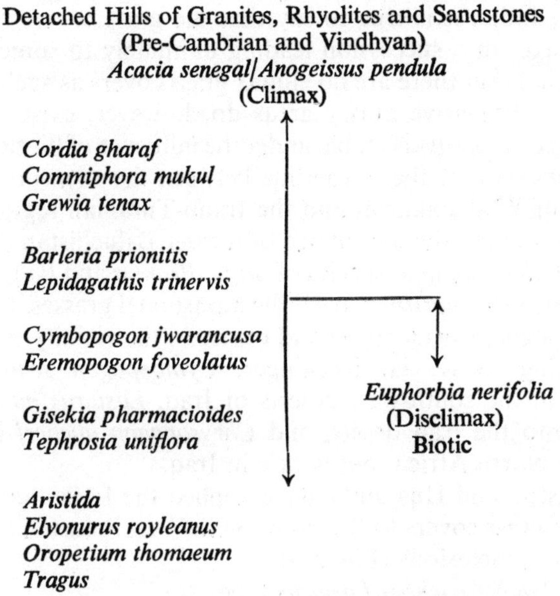

Detached Hills of Granites, Rhyolites and Sandstones
(Pre-Cambrian and Vindhyan)
Acacia senegal/Anogeissus pendula
(Climax)

Cordia gharaf
Commiphora mukul
Grewia tenax

Barleria prionitis
Lepidagathis trinervis

Cymbopogon jwarancusa
Eremopogon foveolatus

Gisekia pharmacioides
Tephrosia uniflora

Aristida
Elyonurus royleanus
Oropetium thomaeum
Tragus

Euphorbia nerifolia
(Disclimax)
Biotic

ston and Ijaz Hussain).[102] Much of the plant cover has been subjected to use and misuse for millennia, and all that remains are various stages of regression of the original cover. H. G. Champion's forest types[30] still provide the basis for classification of the forest vegetation. More recently, Y. I. Selod[199] has applied the Gaussen approach to a study of bioclimates and vegetation of West Pakistan, and recognizes nine bioclimatic regions.

According to the Meteorological Survey of Pakistan, there are three air masses that bring rain to West Pakistan:

(*a*) The Bengal monsoon that originates in the Bay of Bengal in May and June and moves westward along the south slopes of the Himalaya from Bhutan to Afghanistan. This monsoon brings rain to the rich lands of the Punjab.

(*b*) The Arabian monsoon, originating in the Arabian Sea, and affecting only the southern part of West Pakistan.

(*c*) Western disturbance. As the continental system of anticyclones covers West Pakistan there is a long and cold continental winter. This is sometimes disrupted by cyclones or western disturbances that come from Iran and Afghanistan. These disturbances also meet the Hindu Kush and bring precipitation to the north-west part of the ranges.

As in India, it is true to say that the grass covers represent seral stages in a succession leading ultimately to some type of forest, and that there are no climax grass covers as such in West Pakistan. Extensive grassy areas do, however, exist and may remain comparatively stable under the influence of biotic factors.

It appears that the borderline between the monsoonal grass covers of West Pakistan and the Irano-Turanian region to the west runs along the mountains of former Baluchistan (Fig. 33). Annual and perennial species of *Stipa*, *Bromus* and *Poa*, together with *Artemisia maritima*, meet the monsoonal grasses, *Cenchrus*, *Cymbopogon*, *Chrysopogon* and others in the vicinity of Quetta. According to K. H. Rechinger, *Cymbopogon schoenanthus* occurs in the south-west deserts of Iraq, *Lasiurus hirsutus* extends into the Iraq desert, and *Chrysopogon aucheri* is found even in North Africa, but is rare in Iraq.

Johnston and Hussain[102] have applied the Indian concept of types of grass covers to the situation in West Pakistan and draw tentative conclusions (Fig. 34):

Dichanthium/Cenchrus/Lasiurus type

In Pakistan as in India, this type appears to be associated with thorn forest, and is distributed throughout the alluvial basin of the Indus River complex from the former North West Frontier Province through the Punjab to Sind and as far as Baluchistan. It is characteristic of the plains but in the north-west and in Baluchistan is also found on low, eroded hills. Summer temperatures are high, while winter temperatures may occasionally fall below freezing point. Rainfall varies from more than 500 mm. in the north to less than 125 mm. in the south and east.

P. M. Dabadghao[59] has listed the most important grass species: *Dichanthium annulatum*, *Cenchrus ciliaris*, *C. setigerus*, *Lasiurus hirsutus*, *Eleusine compressa*, *Cynodon dactylon*, *Sporobolus marginatus* and *Panicum antidotale*. *Zizyphus nummularia* and *Prosopis spicigera* are important among the shrubs and trees. Five regressive sub-covers have been recognized, namely *Cenchrus*, *Lasiurus*, *Cynodon*, *Eleusine* and *Aristida/Eragrostis*. In West Pakistan, the type appears to be represented by these sub-covers and 'climax' stands are rarely encountered. *Dichan-*

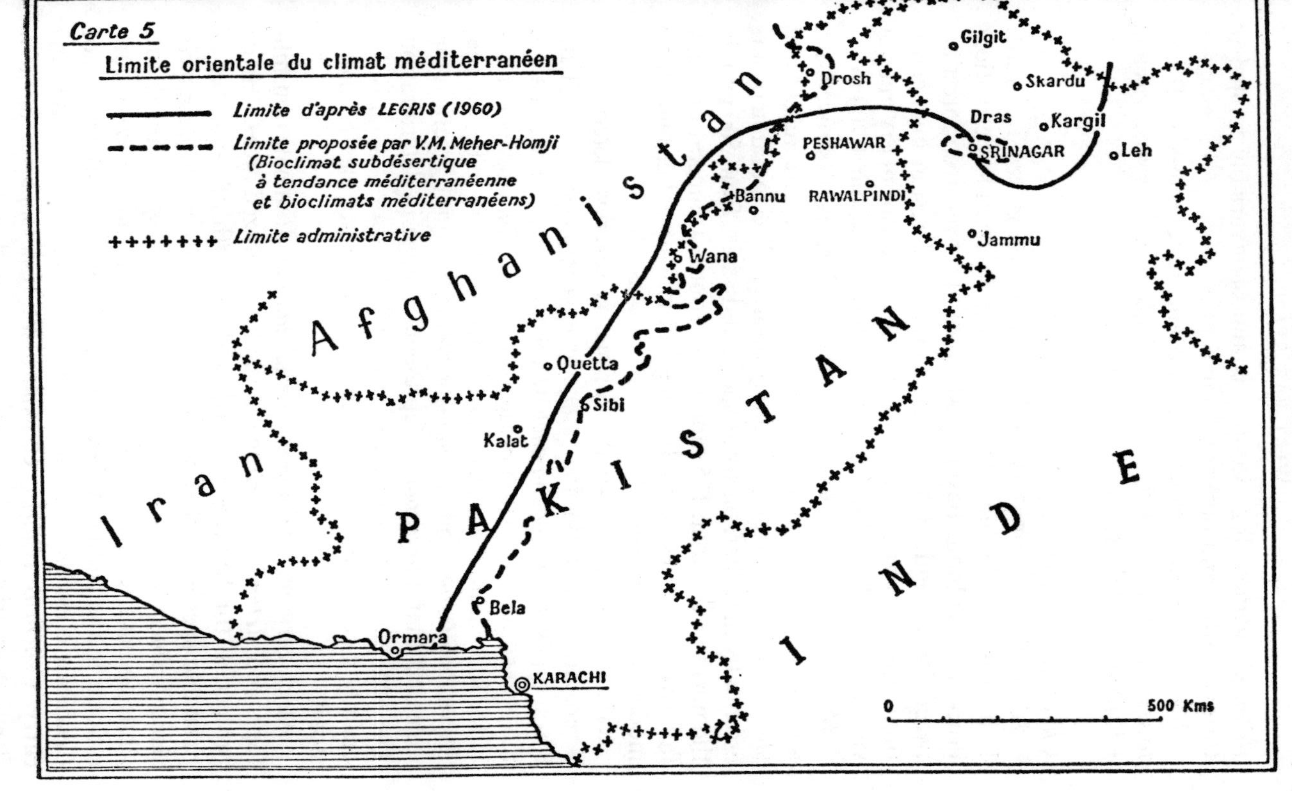

FIGURE 33

Eastern limit of the Mediterranean climate, according to J. Legris and V. M. Meher-Homji[143]

thium annulatum, for example, is almost invariably limited to thorny bushes and other protected spots.

Rangelands near Dabeji, in Kohistan, are composed of *Dactyloctenium sindicum*, *Aristida adscensionis*, *Cenchrus setigerus* and *Chrysopogon fulvus*.

Chrysopogon type

A number of forest types have been recognized in former Baluchistan, including elements of thorn forest at low elevations bordering the Indus Basin, sub-tropical dry evergreen forest between 750 m. and 2,000 m., dry temperate forest at high elevations, and *Artemisia* steppe, which is characteristic of dry interior valleys of the Himalaya, and which starts at about 1,200 m.

Many millenia of overgrazing have largely destroyed the natural vegetation of Baluchistan. The forests are relicts and show little evidence of natural regeneration. Precipitation occurs essentially in winter, partly as snow, and varies from 500 mm. in the north to less than 175 mm. in the south. Temperatures may fall several degrees below freezing in the winter, but tend to be moderate in summer.

Grass vegetation, like the forests, is severely deteriorated. There are indications that *Chrysopogon* spp. occur in all the forest types. *Chrysopogon aucheri* appears to be characteristic of the arid southern portion of the region, and *Chrysopogon fulvus* of the more humid northern part. There is reason to believe that *Stipa*, *Pennisetum* and *Enneapogon* were important genera at one time and that *Chrysopogon* represents a stage in deterioration. The ecological status of *Cymbopogon jwarancusa*, which is frequently a co-dominant with *Chrysopogon*, is not clearly understood.

At Hazarganj, *Cymbopogon schoenanthus* dominates on hilltops, and *Artemisia maritima* in the lower grazed areas. Partial protection for 10 years resulted in the growth of a number of palatable species, including *Chrysopogon aucheri*, *Stipa szowitsiana*, *S. linearis*, *Enneapogon persicus* and *Oryzopsis aequiglumis*, while *Bromus tectorum* was more common under grazing conditions. In the Isplingi valley, on coarse alluvial soils, species of *Cenchrus*, *Cymbopogon* and *Stipa* are found, while gravel fans carry *Cenchrus ciliaris*, *C. panicoides*, *Chrysopogon aucheri*, *Pennisetum orientale*, *Tetrapogon villosus* and *Aristida* spp.

Near Quetta, annual *Poa* and *Bromus* are important in the low-lying areas, while *Chrysopogon* and *Cymbopogon* are found in relative abundance on deeper soils of hill slopes.

Themeda/Arundinella type

Dabadghao[59] has recognized a *Themeda/Arundinella* type which characterizes the northern mountains of India at elevations from 500 to 2,500 m. The principal grass species are: *Themeda anathera*, *Arundinella* spp., *Eulaliopsis binata*, *Chrysopogon* spp., *Dimeria* spp., *Bothriochloa* spp., *Heteropogon contortus* and *Pennisetum orientale*. The following deterioration stages have been recognized: *Dimeria*, *Eulaliopsis*, *Chrysopogon*, *Bothriochloa*, *Heteropogon* and *Eragrostis/Eragrostiella*. In a degraded form, this type is characteristic of the hill pastures in

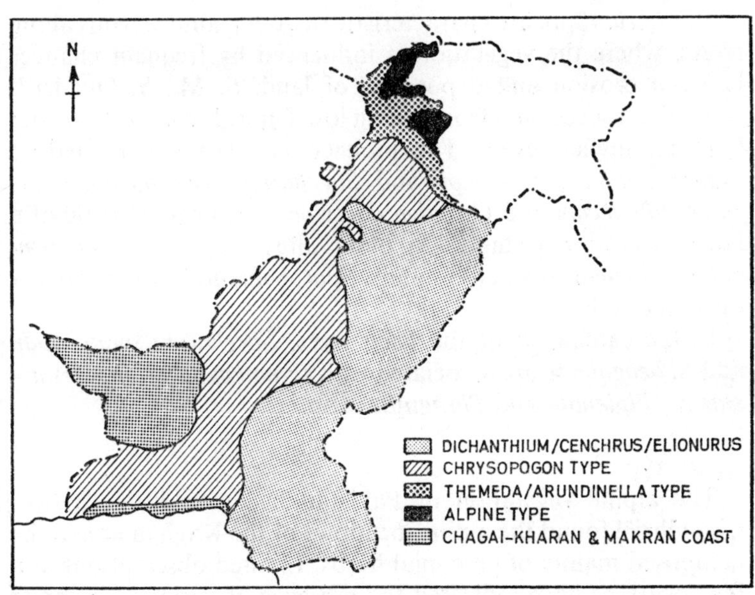

FIGURE 34

Distribution of proposed grass cover types of West Pakistan (*Saccharum* type omitted because it is confined to waterlogged soils, along streams and rivers and beds of nullahs), according to Alexander Johnston and Ijaz Hussain[102]

such areas as Murree, Hazara and Swat; *Heteropogon contortus* is dominant in much of the area, due to heavy summer grazing and probably also to frequent burning.

The type has been little studied in Pakistan. I. I. Chaudhri[34] reports on the vegetation of the Kaghan Valley that the important grass species are *Heteropogon contortus*, *Themeda anathera*, *Oryzopsis monroi*, *Chrysopogon serrulatus* and *Cymbopogon distans*, while the common shrubs in the grasslands are *Indigofera gerardiana*, *Plectranthus rugosus*, *Berberis lycium*, *Daphne oleoides*, *Myrsine africana*, *Punica granatum* and *Asparagus* spp. A. Khan and A. G. Bhatti[118] found *Heteropogon contortus*, *Chrysopogon fulvus*, *Pennisetum orientale* and *Themeda anathera* in the Murree hills.

Saccharum type

Saccharum spp. are characteristic of young alluvial soils along rivers, where the vegetation is influenced by frequent changes between erosion and deposition of land. S. M. A. Quadri[178] studied the succession from young low-lying alluvial soil to older land on higher levels. The pioneer association consisted of *Tamarix dioica*, *T. troupii* and *Saccharum spontaneum*, while *S. arundinaceum* and *Cynodon dactylon* were characteristic of a later successional stage. On older alluvium, *Saccharum bengalense* formed very dense stands. *Sporobolus arabicus* may occur on saline soils.

In the catchment of the Ravi river, *Saccharum spontaneum* and *S. bengalense* are associated with *Cynodon dactylon*, *Desmostachya bipinnata* and *Phragmites* spp.

Alpine type

The alpine vegetation of Pakistan has been little studied. Chaudhri[34] found the alpine meadows of the Kaghan area to be composed mainly of perennial herbs. Limited observations at a few locations have indicated the presence of various temperate species such as *Poa*, *Festuca*, *Bromus* and *Potentilla*.

Chagai-Kharan and Makran Coast

Much of Chagai-Kharan consists of vast areas of black oxidized rocks and pebbles with saline depressions that cannot

36. India. An improved pasture of *Dichanthium annulatum* in Pali District, Rajasthan. Clayey loam. Scattered *Salvadora oleoides, Zizyphus, Capparis aphylla, Prosopis spicigera*. (*Photo Mahendra Prakash*)

37. India. Contour furrowing on sloping lands for pasture improvement in Pali District, Rajasthan. Rocky site. (*Photo Mahendra Prakash*)

38. Taiwan. Vast high mountain meadows on the dry top of TA HSÜEH SHAN (*Indocalamus niitakayamensis* consocies). (*Photo Taiwan Forestry Research Institute*)

39. Taiwan. Hay stand of *Miscanthus transmorrisonensis* in flower. (*Photo Taiwan Forestry Research Institute*)

40. Taiwan. Transition zone of *Miscanthus transmorrisonensis* and *Indocalamus niitakayamensis* (latter in lower right corner). (*Photo Taiwan Forestry Research Institute*)

support vegetation. The Makran coast has been described as a desert waste. The average rainfall is from 50 to 100 mm. per year, and summer temperatures are high. *Haloxylon ammodendron*, a species of medium palatability, occurs throughout these areas.

The *Dichanthium/Cenchrus/Lasiurus* type has the greatest potential as grazing range, with a carrying capacity of 10 hectares per animal unit on good condition range and 20 hectares on fair condition range, although other specialists have quoted 80 hectares per animal unit for desert range. The land under the *Chrysopogon* type has been severely eroded, the fertile topsoil has gone, and native grasses are no longer present to provide a source of seed through which succession may progress from the present degraded condition. Various authorities state that the range lands of West Pakistan are at least three to five times overloaded with free-ranging animals of all types.

TIBET (SITSANG)

No genuine understanding of China is possible without a realization of the fact that the nation is founded on two great natural plant formations, the woodlands in the east and the grassland/desert complex in the west,[241] two vast areas approximately equal in area. In the 37 centuries of recorded history since the Hsia dynasty (2205–1766 B.C.) and in the many centuries before that period, the grasslands and the woodlands have been greatly changed through human activities and this change may be accelerated in the future. The vegetation in its turn has exerted a profound influence on the nation during the historic era and will undoubtedly influence the course of its development in the future. The two great divisions, the woodlands of the east and the grasslands of the west, remain the most influential of the nation's natural environments.

The contrast between the western grassland region of China and the monsoon grasslands of Pakistan, India and Burma is noted here (Table 16). The forest climax regions in eastern China come within the monsoonal transect proper (Chapter VII).

TABLE 16

China. The forest east and the grassland west[241]

The East	The West
1. Essentially woodland, except under unfavourable local habitat conditions.	1. Grassland and desert. Trees are confined to high elevations and areas near water supply.
2. Land cleared primarily for crop cultivation.	2. With the exception of the arable grassland, the primary land use is for grazing.
3. Permanent settlement.	3. Nomadic, except for the arable grassland and the oases.
4. Densely populated except in the undeveloped frontiers and in the montane-boreal regions.	4. Sparsely populated except in the limited irrigated areas.
5. The population is 250 to more than 350 per sq. km. in the densely settled agricultural provinces, and around 100 per sq. km. in the hilly regions.	5. The population is less than 50 per sq. km. in the grassland region and less than 2 per sq. km. in the steppe and desert.
6. Alluvial plains, broad river valleys, basins with external drainage, hills, dissected plateaux, and mountain ranges.	6. High plateaux, basins without external drainage, and high mountain ranges. Essentially an internal drainage area.
7. General elevations below 200 m. in the plains and 500–1,000 m. in the hilly region (except in the Southwestern Provinces).	7. General elevations 1,000–1,500 m. in the steppe and desert regions, and 4,000–5,000 m. in the plateau.
8. Mountain ranges to 3,000 m. The highest peaks rarely exceed 4,000 m. (Taipai-Shan).	8. Mountain ranges 6,000–7,000 m. The highest peak 8,882 m. (Chomolungma).

At elevations around 3,500 m. on the slopes of the Himalaya, the monsoonal grasses and grass communities of India, Nepal and West Pakistan meet the overspill of an entirely different world, the world of *Agropyron*, *Elymus*, *Festuca*, *Poa* and *Stipa*. This is the southern limit of those vast expanses of grassland and desert that extend over the arid heartland of the great Eurasian land mass from the Amur to the Danube. 'They are among the most extensive of the world's vegetation types' (Wang).[241] In

China the grassland and desert occupy about half of the total land area. They cover the immense arid regions of the Chinese interior in a continuous unbroken expanse from the central prairie of the North-eastern Provinces to the mountain steppe of the Pamirs, a distance of 50° of longitude.

Outer China is the domain of the pastoral nomad (Buchanan,[21] Phillips[170]), or the cultivators of the oases. Two main subdivisions may be recognized: the high-level tundra plateau of Tibet (Fig. 35) and the mountain ranges and desertic basins of Central Asia and the Gobi. Over much of Tibet, the general elevation exceeds 3,600 m., the frost-free period is less than 50 days, and in no month does the mean temperature exceed 10° C. Most of the area is a waste of frozen desert with patchy scrub and grass cover.

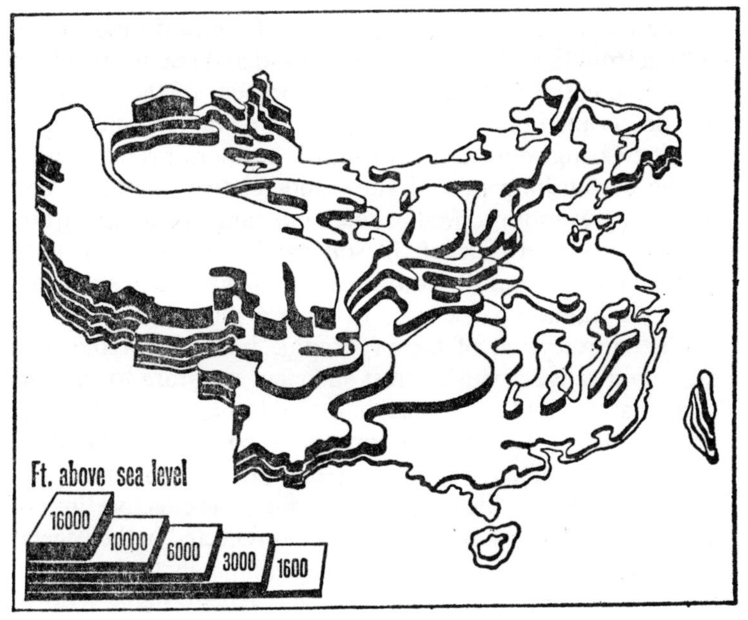

FIGURE 35

Relief map of China, showing in this context the contrast between the east and west, and between the Tibetan Plateau and the hills and ranges to the north, according to Keith Buchanan[21]

North of this plateau, and hence less relevant to the present discussion, are the great mountain chains—Altyn Tagh, Tien Shan and Altai—which enclose the desert basins of the Tarim, Tsaidam and Dzungaria. The chains rise to from 3,600 to 6,600 m.; the basins are below 180 m. The lowlands all show a rather similar concentric arrangement of land types—a central plain of drift sand, an outer belt of better watered, scrub country, and a discontinuous belt of oases receiving their water from glacier-fed streams. The steppe and the desert are the domain of nomads, chiefly Mongols, grazing sheep, horses and camels. New types of animal husbandry have been introduced, with afforestation for shelter, and irrigation to help the farmers to utilize the frost-free season of 5–8 months—agriculturally a developing pioneer fringe area.[21]

According to Wang:[241]

> The increasing continentality toward the arid interior, resulting from the relative position of land and sea, is one of the most influential factors in the development of the grassland and desert and in the spatial differentiation of the minor types. The increasing continentality toward the interior is manifested in the diminishing precipitation and relative humidity, and increasing evaporation, sunshine, and range of temperature, both annual and diurnal. In soil conditions, the increasing continentality is paralleled by a series of changes in soil types from the humid coastal regions toward the arid inland. . . . With the exception of the orographic influence of the high mountain ranges, which arrest sufficient moisture to support rich grassland and even forest in the midst of desert basins, the extent of the steppe, with its continuous cover of steppe grass, indicates the effective limits of penetration of the moisture-bearing wind from the ocean. In the north the effective limit of the maximum monsoon penetration, as reflected in the vegetational development, corresponds approximately to the north–south trending axis of Long-Shan and Holan-Shan. West of this line begin the desert and semi-desert vegetation and the immense inland region of interior drainage. This transition in vegetation is characterized by the increasing importance of the desert shrubs and the reduction of continuous grass cover. However, a fine line of distinction between the steppe and semi-desert does not exist. Steppe and desert merge into each

other in such a wide transitional zone, as in western Suiyuan, that the transitional form is a distinct type in that region.

On the other hand a sharp change between a dry valley and a wet one does occur in the parallel gorges of the dissected Sitsang Plateau. These gorges are deep troughs 2000–3000 m. deep. This sharp change reflects the limit of effective penetration of the monsoon in the valley, and a shift in the delicate balance between the moisture bearing monsoon and the desiccating effect of foehn winds from the upper slopes of the lofty plateau.[241]

The history of the Tibetan flora in relation to that of the Himalaya and as part of the Sino-Himalayan botanical region has been noted under India above (p. 166).

The plateau of Tibet has a threefold nature:

Interior plateau with internal drainage—tundra climax. Wang[241] states that the common grasses are: *Agropyron thoroldianum*, *Avena subspicata*, *Elymus junceus*, *E. lanuginosus*, *Festuca valesiaca* (alt. 5,100 m.), *Glyceria distans*, *Poa alpina*, *Stipa purpurea*, *S. sabulosa*.

TABLE 17

Maximum recorded altitudes of species in the Himalaya and Tibet[88, 241]

Species	Metres	Species	Metres
Agropyron longe-aristatum	4,830	*Festuca sibirica*	4,500
A. striatum	4,950	*F. valesiaca*	5,160
A. thoroldianum	4,950	*Glyceria distans* var. *convoluta*	4,860
Allium semenovii	5,100		
A. senescens	4,500	*Littledalea tibetica*	4,950
Arenaria sp.	6,000	*Microula tibetica*	5,400
Aster boweri	5,100–5,400	*Oryzopsis lateralis*	4,200
Astragalus confertus	5,100–5,400	*Oxytropis tatarica*	5,100
A. heydei	5,100	*Poa attenuata*	5,400
Avena subspicata	5,550	*P. hirtiglumis*	5,400
Capsella thomsonii	5,400	*P. nemoralis*	5,100
Cheiranthus himalayensis	5,190	*P. pratensis*	4,500
Cochlearia scapiflora	5,100	*P. tibetica*	4,500
Deschampsia caespitosa	4,950	*Saussurea tridactyla*	5,700
Deyeuxia compacta	4,500	*Stipa hookeri*	4,440
Diplachne thoroldii	4,740	*S. orientalis*	4,500
Elymus dasystachys	5,100	*S. purpurea*	4,500
E. sibiricus	5,400	*S. sibirica*	4,500
Festuca deasyi	3,000	*Thermopsis inflata*	5,100–5,400
F. nitidula	4,500	*Thylacospermum rupifragum*	5,100

Outer plateau, comprising the headwaters of the outward-flowing rivers, Hwang-Ho, Yangtze, Lantsang-Kiang (Mekong), Nu-Kiang (Salween), Chiu-Kiang (Irrawaddy), Trangpo (Brahmaputra), Sutlej and Indus—alpine climax.

River gorge region or dissected plateau, where those rivers are forcing their way through the boundary ranges of the plateau; gorges show two cycles of erosion, ice followed by water—forest climax.

A list of highest recordings of species is compiled (Table 17) from W. B. Hemsley and H. H. W. Pearson[88] and Wang;[241] the latter concludes that there seems to be no upper limit except perpetual snow for the altitudinal distribution of seed plants on the earth's surface. According to Wang, the same genera and often the same species attain the upper limits of the phanerogamic vegetation both in Tibet and the Himalaya. Hemsley and Pearson[88] have shown that at least 29 seed plants are common to Tibet and the European Alps (Table 18).

TABLE 18

Plants common to Tibet and the European Alps[88]

Androsace chamaejasme	*Poa alpina*
A. villosa	*P. nemoralis*
Avena subspicata	*P. pratensis*
Carex incurva	*Polygonum viviparum*
C. rigida	*Potentilla fruticosa*
C. ustulata	*P. multifida*
Clematis alpina	*P. nivea*
Draba alpina	*Salix lapponum*
D. fladnizensis	*Saussurea alpina*
Festuca valesiaca	*Saxifraga hirculus*
Gentiana tenella	*Scirpus caricis*
Leontopodium alpinum	*Sedum rhodiola*
Oxytropis lapponica	*Taraxacum officinale*
Pedicularis oederii	*T. palustre*
	Thalictrum alpinum

NEPAL

An altitudinal zonation of the vegetation of Central Nepal has been proposed by members of the Japanese expedition organized in 1952–3 by the Fauna and Flora Research Society of Kyoto University (S. Kitamura).[120]

NEPAL

Altitude in metres	Vegetation	Zone
5,200	Permanent ice and snow	
4,000 to 5,200	Alpine shrubs and herbs	Arctic
3,200 to 4,000	Conifers (*Abies*) and *Betula*	Cold temperate
2,500 to 3,200	Conifers (*Tsuga, Picea*) and deciduous trees	
2,000 to 2,500	Evergreen trees (*Quercus glauca*, etc.)	Warm temperate
1,200 to 2,000	*Myrica, Photinia* etc.	
1,000 to 1,200	*Castanopsis indica*	
600 to 1,000	*Shorea robusta*	Subtropical

TABLE 19

Gramineae of Nepal recorded by J. Ohwi

	Altitude of collection, in metres
Agropyron longe-aristatum	2,600, 4,400
Arundinella hookeri	2,800, 3,100
Avena fatua (barley and wheat fields)	2,900
Brachypodium sylvaticum var. *luzoniense*	3,400
Chrysopogon gryllus	2,200
Cynodon dactylon	2,900
Digitaria adscendens	680, 700
Eleusine coracana	1,700
Festuca leptopogon	2,100
F. polycolea	4,500
Helictotrichon sp.	4,000–4,500
Imperata cylindrica	1,500
Isachne albens	2,000
Koeleria cristata	3,300, 3,500
Paspalum distichum	1,400
Pennisetum alopecuros	2,600
Poa himalayana	3,700
P. pratensis	2,600, 2,900
P. tibetica	2,900, 3,000
Pogonatherum crinitum	1,400, 1,800
Setaria geniculata	1,650
Stipa spp.	2,900
Trisetum spicatum	3,300

Thus again in this border zone we find the Indian subtropical and tropical grasses at the lower altitudes, and the grasses of the cold lands of Asia to the north spilling over at the higher altitudes. In his ecological remarks, Sasuke Nakao notes how at an altitude of 4,000–4,500 m. on the northern flank of Annapurna, one grass (a species of *Helictotrichon*) dominated the vegetation, with other herbs or sedges rare. This kind of knee-high grassy vegetation occurs just above the timber line, a grassy zone constituting an upper series of the thorny bush zone, with *Betula utilis* and *Caragana gerardiana*. This climax type of alpine grassland is peculiar because of the absence of the beautiful flowers that characterize this zone in the Himalaya. In the alpine and subalpine zones, the selective grazing of yaks, cows and their hybrid, zhos, favours the survival of *Primula*, *Rhododendron*, *Meconopsis*, *Berberis* and *Iris*. Intensive grazing of narrow hills around villages at the borders of Kathmandu Valley creates a low turf composed of *Pogonatherum crinitum*, *Paspalum distichum* and *Imperata cylindrica* var. *major*.

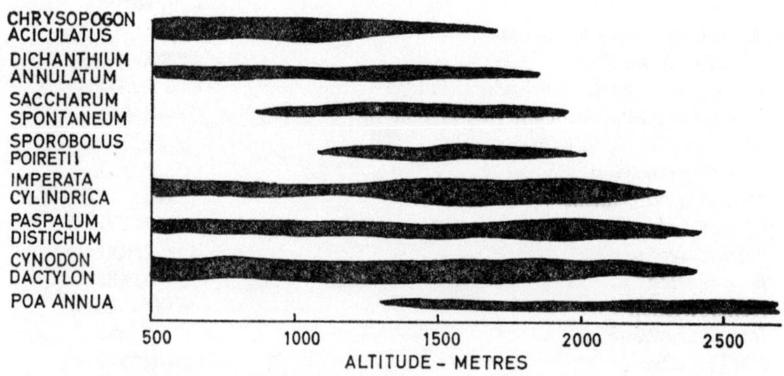

FIGURE 36

Altitudinal distribution of grass species in eastern Nepal according to M. Numata[158]

M. Numata[158] has also provided information on altitudinal zonation and on parallelism between forest type and ground cover (Table 20). In the subtropical lateritic zone, many grasses grow well in the rainy season, e.g. *Chrysopogon*, *Saccharum*, *Dichanthium annulatum*, *Cenchrus ciliaris*, *Heteropogon contortus*, *Phalaris minor* and *Bothriochloa pertusa*. The grass cover type dominated by *Cynodon dactylon* and *Imperata cylindrica* in the subtropical, warm temperate and temperate ecotones disappears above 2,500 m. and the cover characterized by *Carex* begins, associated with species of *Festuca*, *Poa*, *Arundinella*,

Table 20

Climatic and vegetation zones in eastern Nepal[158]

<table>
<tr><th>Climatic zone</th><th>Altitude (m.)</th><th>Climax forest</th><th>Secondary forest</th><th>Grasslands including dwarf bamboo</th></tr>
<tr><td>Arctic</td><td>3,800–5,000</td><td>Juniperus
Rhododendron</td><td></td><td>Helictotrichon, Primula, Androsace, Iris, Berberis, Saussurea</td></tr>
<tr><td>Subarctic</td><td>2,900–3,800</td><td>Picea
Rhododendron</td><td></td><td rowspan="2">Arundinaria, Carex, Arundinella, Eragrostis, Deschampsia, Festuca, Primula, Androsace, Potentilla, Artemisia, Pieris</td></tr>
<tr><td>Cool temperate</td><td>2,500–2,900</td><td>Tsuga
Rhododendron</td><td rowspan="2">Pinus excelsa</td></tr>
<tr><td>Temperate ecotone</td><td>1,900–2,500</td><td>Quercus
Cinnamomum
Camellia</td><td rowspan="2">Imperata cylindrica, Cynodon dactylon, Arundinaria hookeriana, Paspalum distichum</td></tr>
<tr><td>Warm temperate</td><td>1,200–1,900</td><td>Schima
Castanopsis
Lithocarpus</td><td rowspan="2">Pinus longifolia</td></tr>
<tr><td>Subtropical</td><td>200–1,200</td><td>Shorea
Bombax
Ficus</td><td>Saccharum, Dichanthium, Chrysopogon, Cenchrus, Imperata, Cynodon, Heteropogon</td></tr>
</table>

Brachypodium, *Eragrostis*, *Iris*, *Primula*, *Androsace*, *Gentiana*, *Artemisia*, etc. Grasslands above 2,500 m. are low in productivity and seriously overgrazed by excessive numbers of livestock. Indicator plants of overgrazing include *Primula*, *Androsace*, *Euphorbia longifolia*, *Artemisia*, *Potentilla*, *Gentiana* and *Plantago* (see Fig. 37).

FIGURE 37

Succession in overgrazed grassland in the cool temperate zone in eastern Nepal[158]

wet dry	*Juncus-Ranunculus* *Primula* *Plantago-Hydrocotyle* *Artemisia* *Potentilla*	*Rhododendron* *Pieris—Pinus excelsa*	 *Tsuga-Rhododendron-Arundinaria*
	Overgrazed type of grassland	Seral stage	Climax

BURMA

This country extends over more than 18° of latitude, from sea level to perpetual snow at over 5,700 m., and with rainfall varying from under 600 to over 6,000 mm. per annum. Under these most diverse ecological conditions, it is natural that there should be a great variation in the flora. According to D. Rhind,[186] there is no comprehensive flora of Burma, nor even an enumeration of the plants comparable to those of most of the surrounding regions. He has listed the known species of the Gramineae, for possible use when such a flora may be written and to preserve such information as exists (Table 21). Many scientific data were lost at the time of the Japanese invasion of Burma, but an attempt has been made to preserve as much as possible of what was saved (see also Bor[19]).

Past work on the grasses of Burma has been almost entirely due to the efforts of the numerous Forest Officers who have been almost the only collectors of plants in the country. In a country the life and well-being of which is intimately bound up with its crops and vegetation, it is remarkable how little has been done to study its plant life systematically. There is, in particular, a noticeable lack of collection from places of little direct economic interest such as, for example, the Chin Hills and the eastern Shan States.

Burma is a land 'wreathed in bamboos'. The whole existence of the Burman is bound up with bamboos. These play a very important role in his everyday life, providing his house, much of his furniture, utensils, farm implements, baskets, containers, binding materials and some of his food. The rural Burman cannot imagine an existence in a country without bamboos. Often the countryside is more effectively dominated by the bamboos than by the larger trees. Certain species are confined to particular habitats. Since different species have specific uses, there is much internal trade in cut bamboos which are transported about the country in large quantities.

Naturally, because of the wide range of ecological conditions, there is a large number of species of grasses in the country, but the list is incomplete. It is regrettable that so few species from high altitudes have been collected, and that with few exceptions the collections represent the grass flora of the more accessible parts of the country.

TABLE 21

The Grasses of Burma[186]

Southern rain forests:

- *Acroceras* spp.
- *Alloteropsis* spp.
- *Coix* spp.
- *Centotheca* sp.
- *Imperata* spp.
- *Microstegium* spp.
- *Oplismenus* spp.
- *Saccharum* spp.
- *Sclerostachya* spp.
- *Sorghum* spp.

Deltaic monsoon areas:

- *Axonopus* spp.
- *Chrysopogon aciculatus*
- *Cynodon* spp.
- *Dichanthium annulatum*
- *D. caricosum*
- *Echinochloa crusgalli*
- *E. stagnina*
- *Eragrostis unioloides*
- *Hemarthria compressa*
- *Isachne albens*
- *I. globosa*
- *Ischaemum* spp.
- *Leersia* spp.
- *Neyraudia* spp.
- *Ottochloa* sp.
- *Paspalidium* spp.
- *Phragmites* spp.
- *Rottboellia exaltata*
- *Saccharum* spp.

Northern wet zone:

Resembles Lower Burma rain forest in grass flora, but *Phragmites* less common and *Coix* more so.

Alloteropsis spp.
Arthraxon spp.
Bothriochloa intermedia
Elytrophorus sp.
Eragrostis unioloides
Eulalia spp.
Isachne globosa
Leersia spp.
Panicum auritum
Sacciolepis spp.
Themeda spp.

Dry zone (600 to 1000 mm. rainfall):

Aristida spp.
A. depressa
Bothriochloa pertusa
Chloris barbata
Cymbopogon spp.
Echinochloa crusgalli
E. stagnina
Eragrostis spp.
Heteropogon contortus
Oropetium thomaeum
Perotis indica
Ratzeburgia sp.
Saccharum spontaneum
Setaria glauca
Sporobolus coromandelianus
S. tremulus
Themeda spp.
Tragus biflorus
Vetiveria zizanioides

Hills (1,200 to 1,800 m.) (1500 to 2500 mm.):

Arthraxon lancifolius
Arundinella spp.
Chrysopogon aciculatus
Coix spp.
Eragrostis tenuifolia
Erianthus spp.
Eulalia spp.
Imperata spp.
Microchloa indica
Microstegium nudum
Muhlenbergia huegelii
Panicum watense
Paspalum scrobiculatum
Pennisetum alopecuroides
Saccharum spontaneum
Sporobolus indicus
Themeda spp.
Thysanolaena spp.

In the southern rain forests, the species tend to be large-leaved, weak-stemmed shade lovers. In the deltaic monsoon areas, some of the species withstand flooding, others do not. Some of the species listed here are characteristic of paddy field bunds and low uplands. The northern wet zone extends from about Kawlin to the north of Fort Hertz. Between the northern and southern wet zones lies the dry zone with a very different flora—an area with a rainfall of 600–1,000 mm. all falling in the monsoon season between May and November, with hot dry weather for the rest of the year. In the hills the grass flora is of two main types: woodland and open downland. The woodland is mostly a fairly open oak-chestnut association with light undergrowth so that shade species of grasses are rare. See also A. McKerral.[140]

CEYLON

The principal factors of the climatic complex are:

(*a*) Rainfall, particularly its distribution rather than the total amount during the year.

(*b*) Atmospheric humidity, the most significant property of which is its saturation deficit.

(*c*) Temperature.

(*d*) Wind.

(*e*) Intensity of insolation.

All these are influenced and considerably modified by the physiographic features (C. H. Holmes).[91] There are two major climatic zones, popularly referred to as the wet and dry zones. The former comprises the south-western quarter extending from the south-western seaboard to beyond the central mountain massif, the rest of the country occupied by the remainder of the first peneplain constitutes the dry zone. The wet zone receives both the south-west and the north-east monsoons and the mean annual rainfall varies from 1,800 to 6,000 mm. per annum. The main rainfall of the dry zone comes from the north-east monsoon, varies from 600 to 2,500 mm., but is mainly over 1,200 mm. per annum.

Seven major forms of vegetation of other than closed-forest type, composed of savannah, grass and farmlands, are found, according to C. H. Holmes,[91] upon whose account of the grassland types of Ceylon the following sections have been based:

A. *Wet zone*

(i) Low country below 450 m. (1) *Talawa* grasslands, (2) Kekilla farmlands.

(ii) Mid-country from 450 to 1,500 m. (3) Dry patana grasslands, (4) Savannah (sub-Himalayan type).

(iii) Hill country over 1,500 m. (5) Wet, black patana grassland.

B. *Dry zone*

(i) Low country. (6) Dry *damana* grasslands, (7) Wet *villu* grassland.

TALAWA GRASSLANDS

This type occurs in the wet low country of the south-western quarter of the island, on the lower slopes of the outermost foothills of the central mountain mass and on the slopes of low sub-coastal hills, between remnants of tropical wet evergreen forest (rain forest) and the cultivated lands in the valleys (coconut, village gardens and paddy). The grass sward is burnt periodically once or twice in the year, but parts may escape burning for periods of a few years at a time. Scattered trees occur in clumps, and the common shrubs and herbs also tend to form clumps. The rainfall varies from 2,500 to 5,000 mm. per annum, quite evenly distributed, but with peaks in May and November and the driest months in February and March. Homoclimes occur in the Andaman Islands, most of Malaysia, the Western Ghats of India and northern and southern Burma, where the climax vegetation is again tropical wet evergreen or rain forest. The soil in the Talawa grasslands is generally an impoverished, truncated, lateritic gravel.

The principal grasses are *Chrysopogon* spp., *Cynodon dactylon*, *Ischaemum ciliare*, *Themeda tremula* and *Dimeria lehmannii*. Occasional patches of *Imperata cylindrica*, *Cymbopogon confertiflorus** and *Themeda tremula* appear to be confined to more fertile lower slopes; *Eriocaulon sexangulare* is common in moist situations. On the dry impoverished soil of the top slopes, *Massia triseta* forms a rather scanty cover in association with sedges. The fern, *Gleichenia linearis*, is found in pockets of deep soil or silt; everywhere there are the dodder-like *Cassytha filiformis* and clumps of *Nepenthes distillatoria*. Associated trees are *Careya arborea*, *Madhuca fulva* and *Calophyllum calaba*; shrubs are *Syzygium caryophyllatum*, *Hedyotis fruticosa*, *Osbeckia aspera* and *Salacia reticulata*.

The development from the grass to the scrub stage and so to forest is dependent upon freedom from fire for a sufficient length of time. The scrub stage is actually more inflammable than the grass stage, and damage ceases only when moister microclimatic conditions supervene as a result of continued close cover of the ground, accumulation of litter, and the building up of soil of better texture and greater moisture retention. So arises the sharp delineation characteristic of the boundaries between grassland

* *Cymbopogon confertiflorus* of Ceylon = *C. nardus* var. *confertiflorus* which, according to Bor,[19] is at home in the mountains of Madras, but was introduced into Ceylon and Malaysia. *C. nardus* var. *nardus* is cultivated widely in Ceylon for citronella oil.

and the stages of seral succession leading back to the closed forest climax.

DRY PATANA GRASSLANDS

These 'grasslands of the hills' are the largest in extent and the most widespread, but are confined to the mid-hill country from 450 to 1,500 m. elevation. They are in four geographical groups. Two are outside the south-central mountain region, one in the highlands between Deniyaya and Rakwana and the other in the Knuckles Range. The other two are in the central mountain zone, one occurring as isolated patches in a 'sea' of tea plantations in the western basin, the other comprising the extensive, continuous and best-known patanas of the eastern Uva basin. These grasslands cover hill slopes of varying gradients from flattish plateaux to steep slopes and escarpments between the second and third peneplains. The physiognomic features are uniform despite floristic differences. All are edaphically dry hill grasslands existing under relatively wet climatic conditions. Forests and grasslands occur together, each sharply defined and separated from the other without any appreciable transitional ecotones between them. The forests are all of a wet evergreen type varying from low-country tropical wet evergreen forest to subtropical wet evergreen montane forest. The grasslands have a dry surface in all weathers and carry a sward of coarse dry tussock-forming grasses rather than spreading or carpet grasses. Trees occur singly, clumps are rare, but shrubs may be present in extensive closed patches covering hillsides. Burning is widespread as in talawa grasslands.

The mean annual rainfall varies from 1,250 to 1,800 mm. in the eastern basin to over 5,000 mm. in the western section of the western basin and part of the Knuckles Range. Mean monthly temperatures are around 22° C, perhaps because of clouds and mists. All four regions are subject to intense insolation during dry clear days, the temperature of the surface soil, particularly after a burn, may be excessively high, and the conditions of the atmosphere favour intense desiccation. Hence the natural vegetation, even of the forest, exhibits many markedly xerophytic characteristics. The soils are seriously impoverished, often truncated clayey loams intermixed with quartz and ferruginous gravel. The hare is an even more pernicious biotic factor than the village cattle.

The sward in the Deniyaya-Rakwana group is composed mainly of *Cymbopogon nardus* var. *confertiflorus* with the slender trailing grass *Pseudanthistiria umbellata* on the higher slopes.

The main element in the grass cover of the Knuckles group is again *Cymbopogon nardus* var. *confertiflorus*, together with *Digitaria thwaitesii* and *Ischaemum aristatum.*

The sward of the Western Basin grasslands is composed mainly of *Cymbopogon nardus* var. *confertiflorus* with some *Pollinia argentea* and patches of *Imperata cylindrica.*

In the Eastern (Uva) Basin there is an almost continuous expanse of grasslands ranging from 900 to 1,250 m. elevation and rising to over 1,500 m. The mean annual rainfall is 1,800–2,500 mm., contributed mainly by the north-east monsoon. There is a relatively dry period between June and August when strong dry winds are prevalent. Mean monthly temperatures are around 20° C from December to February, and rise gradually to 22° C from June to August. Insolation is intense in the open on clear days and this, combined with the high winds, has an extremely desiccating effect on the vegetation.

The grasses vary in frequency, being more luxuriant and tufted on lower slopes and patches of deeper soil, scanty or even absent on the poorest soils of crests and ridges. Nearly all grasses are coarse and wiry; principal species are *Arundinella villosa, Pollinia phaeothrix, Ischaemum ciliare, Aristida setacea, Panicum trigonum, Cymbopogon nardus* var. *confertiflorus* and occasionally *Imperata cylindrica. Themeda tremula* is abundant in sections on better soils, sedges such as species of *Fimbristylis* occur in moist situations as well as in dry patches of poor soil or hill tops. *Pteridium aquilinum* is nearly ubiquitous. *Chrysopogon zeylanicus* which is more characteristic of the wet patana becomes dominant on the higher reaches of the dry patana along with other characteristic species of herbs, shrubs and *Rhododendron.*

Holmes[91] does not agree that the Uva patana grasslands have always been grassland or even savannah. 'The simple truth that forest and grassland do exist under the same climatic conditions merely means that the original state must have been all forest, for no school of ecology can admit of two widely divergent types of climatic climaxes in the same place . . . the only possible conclusion that the original vegetation type within this region must have been tropical wet evergreen forest. . . . After the many centuries that must have elapsed since the original forests of the upper and middle Uva were cleared, it is not unreasonable to expect that under the conditions obtaining there would remain today much direct vegetational evidence.' But John Davy reported in 1821 that *Cymbopogon nardus* var. *confertiflorus* was nearly as characteristic of the Uva patana then as it is now in the dry patanas of proved biotic origin in comparatively recent times. Holmes describes the stages of succession following protection from grazing and burning: 'From whatever angle the dry patana grasslands are looked at—whether climatologically, floristically, ecologically, historically or archaeologically—they are undoubtedly of secondary origin, entirely man-made and so maintained.'[91]

SAVANNAH FOREST

This type is a narrow belt at elevations around 600 m. and down to 300 m. between the dry patana grasslands of the south-central mountain core and the dry evergreen forests of the surrounding plain. It appears to be restricted almost entirely to the south-eastern and eastern limits of the second peneplain. The changes from patana to savannah and from savannah to closed dry evergreen forest are gradual and not abrupt. In the savannah forest itself, the coarse tussock grasses, *Imperata cylindrica*, *Cymbopogon nardus* var. *confertiflorus*, *Anthistiria argeas* and *Sorghum nitidum* cover the ground in a dense layer. They are burnt periodically by cattle graziers and probably also hunters to drive out game. 'There is no reason whatever to postulate that savannah forest of this type did at any time, even as a passing seral phase, cover the whole of the upper Uva grasslands . . . it is merely secondary seral forest composed of fire-resistant tree species associated with inflammable grasses.' It should not be regarded as equivalent to the sub-Himalayan forest tracts.[91]

WET BLACK PATANA GRASSLANDS

These occur, usually closely hedged in by subtropical wet montane evergreen forest, at elevations of generally over 1,500 m. in the region of the central mountain zone that is served by both monsoons. They are entirely restricted to the third and highest peneplain and distributed along the central stem of the mountain anchor, separating the western and eastern basins;

they occupy high plateaux or slopes of easy gradients. The rainfall is relatively high, evenly distributed and provided by both monsoons as well as by pronounced thunderstorms. There is no dry month. Solar radiation is low, due to overcast skies and mists. Cold strong winds are characteristic, nearly always moist and saturated. Mean monthly temperatures vary from 13° to 17° C. Ground frosts are common on clear nights from late December to about the end of February. Thornthwaite's classification would be wet, meso-thermal with rain throughout the year, and Champion's forest type[30] is subtropical montane evergreen type 7(*a*)C.

Unlike the rather uniform surface of the typical dry patana, the surface here is rough and broken by clods and little cushions of tussock grasses. *Chrysopogon zeylanicus* is the characteristic grass, with *Cymbopogon nardus* var. *confertiflorus* in the narrow fringing ecotones between the forest grasslands and in successional stages leading to forests. Other important species are *Arundinella villosa*, *Pollinia phaeothrix* and *Ischaemum ciliare*. The only natural tree species in the black patana are stunted gnarled trees of *Rhododendron arboreum*, festooned with the lichen *Usnea barbata*. Herbs are more common than in the dry pastures and include a number of European forms (*Potentilla*, *Alchemilla*, *Thalictrum*, *Ranunculus*). Wet low-lying areas carry dense masses of the bamboo, *Arundinaria densifolia*; *Pteridium aquilinum* is as frequent and universal as in the dry patana.

In contrast to the dry patana grasslands, it has generally been accepted that the wet black patana grasslands are a secondary seral stage, maintained in that condition by fire and grazing. Reference should be made to the difference of opinions with regard to the ecological status of the similar formation found in the Nilgiris of south India. Wherever protection from fire is given to the wet patanas in Ceylon, changes towards closed forest conditions take place—the *Rhododendron* increases in frequency and forms thickets in which shrubs and trees become established.

DRY DAMANA AND WET VILLU GRASSLANDS

These two types have the same geographical position, and occur in the same climatic zone; the differences between them are due entirely to soil conditions. Both are found throughout the low country dry zone in scattered patches of varying size separated by patches or strips of dry mixed evergreen forest of varying quality and secondary chena (= taungya) seral jungle. They occur in the first and lowest peneplain mostly below 150 m. elevation. The mean annual rainfall is from 1,800 to 2,500 mm.; the driest months, June to August, coincide with the south-west monsoon which changes in the rain shadow to the east of the south central massif as dry hot winds, known as kachan. Mean monthly temperatures range from a minimum of 25° C in January to a maximum of 30° C in June. The forest type is Champion's dry mixed evergreen high forest.[30] There is sharp demarcation between forest and grassland; the grasses are burnt periodically in patches, but not entirely through an area at any one time or in any one year. The damana grasslands are of the savannah type, but with no uniformity in size or distribution of the tree components. There is a vigorous growth of coarse, tufted or tussock-forming grasses such as *Imperata cylindrica*, *Cymbopogon nardus* var. *confertiflorus*, *Chrysopogon aciculatus*, *Ch. montanus*, *Chloris barbata*, *Dactyloctenium aegyptium*, *Aristida setacea* and *Digitaria marginata*.

There is a subtype intermediate between damana and villu grasslands which occurs in areas subject to periodic flooding during the rains. The surface is broken into rounded and raised 'dumpy' grassy cushions separated from one another by channels of alternate flooding and drainage. The trees are less frequent than in damana but more frequent than in villu.

The wet villu grasslands are found in permanently moist sites around ancient tanks, large waterholes or by river banks. The damana type gradually gives place to the villu type with its fresh, more succulent and creeping rather than tufted tussock grasses. At the water's edge occur the semi-floating grasses, *Paspalidium geminatum* and *Oryza fatua*. Further from the water, *Iseilema laxum*, *Paspalidium flavidum* and *Cynodon dactylon* are common to abundant, along with *Stenotaphrum dimidiatum*, under-shrubs, occasional trees and *Fimbristylis argentea*. *Alysicarpus vaginalis* is common.

The dry damana grassland results from the colonization of grasses, principally *Imperata cylindrica*, in areas where dry mixed evergreen high forest has been cleared, burnt and cultivated for

2–3 years under chena crops and then subjected to periodic fires. The oldest dry damana grasslands are believed to have been paddy lands during the heyday of the ancient Sinhalese civilization that flourished and decayed in those parts.

Chapter VII

CHINA (MAINLAND)

Although Keith Buchanan has said that geographers do not distinguish between monsoonal rainfall and summer rainfall, with certain exceptions, it is believed that in the present context we should include much of eastern China within the monsoonal ecoclimate. Because the Tsinling and the Nanling are not so high as the Himalaya, the northern limit of the monsoonal climate is more difficult to define. Perhaps we should say that temperature is here a more important factor than rainfall. In the south we have monsoonal conditions in both summer and winter. As one goes north over the Nanling and then over the Tsinling, the summers may remain monsoonal but the winters become colder due to the influence of northern air streams from Siberia.

The distribution of some of the major crops may indicate which provinces should be regarded as having a monsoonal ecoclimate:

Yangtze rice/wheat:	Anhwei, Chekiang, Hunan, Hupei, Kiangsi, Kiangsu
Rice/tea:	Anhwei, Chekiang, Fukien, Hunan, Kiangsi
Szechwan rice:	Shensi, Szechwan
Double cropping rice:	Fukien, Kwangsi, Kwangtung
South-western rice:	Kweichow, Yunnan

North of the Tsinling divide one finds open vegetation, grasses and bushes, sparse types of grassland and steppe, grading into desert in Inner Mongolia.[159] The Tsinling divide shows in its vegetation the most northerly extension of southern forms. The combined influence of a richer vegetation and higher rainfall causes podsolization. An evergreen broad-leaved forest is characteristic of central and south China, with an increasing proportion of subtropical forms towards the south. Between the Yangtze basin with its highly mixed type of vegetation, and the

1 MONTANE CONIFEROUS FOREST PREDOMINATED BY SPRUCE AND FIR
2 MONTANE CONIFEROUS FOREST PREDOMINATED BY LARCH
3 MONTANE CONIFEROUS FOREST PREDOMINATED BY SPRUCE
4 MONTANE CONIFEROUS FOREST PREDOMINATED BY FIR
5 MIXED NORTHERN HARDWOOD FOREST PREDOMINATED BY BIRCH
6 MIXED NORTHERN HARDWOOD FOREST OF MAPLE, BASSWOOD AND BIRCH
7 DECIDUOUS BROAD-LEAVED FOREST PREDOMINATED BY DECIDUOUS OAKS
8 MIXED MESOPHYTIC FOREST
9 EVERGREEN BROAD-LEAVED FOREST OF EVERGREEN OAKS, SCHIMA & LAURELS
10 EVERGREEN BROAD-LEAVED FOREST OF EVERGREEN OAKS, SCHIMA AND LAURELS WITH PINUS MASSONIANA IN SECONDARY STANDS
11 EVERGREEN BROAD-LEAVED FOREST OF EVERGREEN OAKS, SCHIMA AND LAURELS WITH PINUS YUNNANENSIS IN SECONDARY STANDS
12 RAIN FOREST
13 GRASSLAND
14 DESERT VEGETATION
15 SALINE VEGETATION
16 TUNDRA AND ALPINE VEGETATION
0 400 800 km

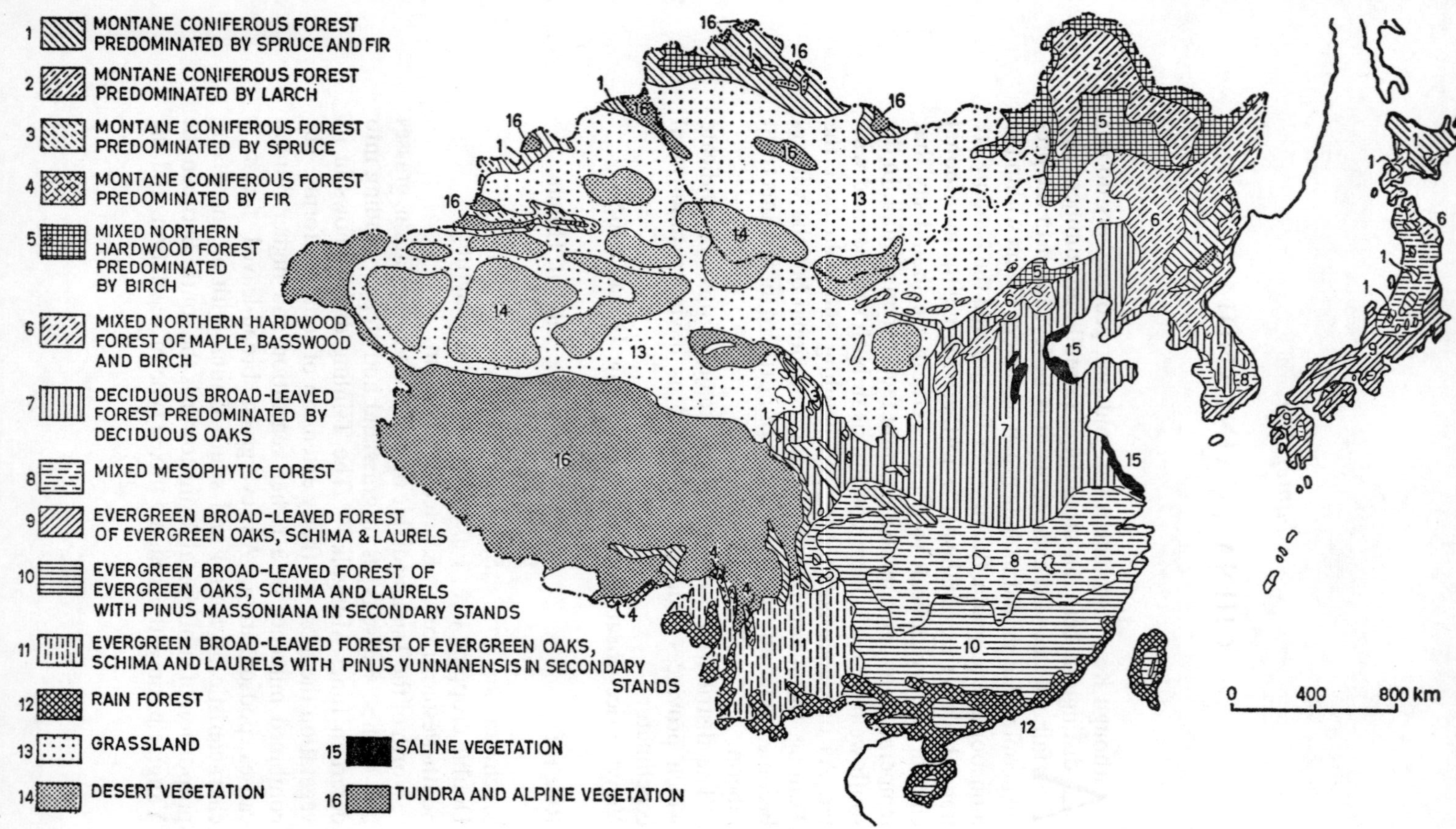

FIGURE 38

dominantly subtropical Sikiang valley, the Nanling forms a vegetational divide comparable with the Tsinling to the north.[241]

The Chinese People's Republic Ministry of Agriculture arranged for the following centres to collaborate in the publication in 1958 of an *Illustrated Manual of Forage Plants of China*: Institute of Plant Breeding and Cultivation and Institute of Animal Husbandry of the Chinese Academy of Agricultural Sciences, Peking Agricultural College, Department of Horticulture of the former North China Institute of Agricultural Sciences, and Popular Science Publishers. An English translation of the manual has been made available by the U.S. Department of Commerce.[38] The publication contains illustrations and descriptions of some 500 species of wild and cultivated plants that are considered to have some forage value. Each item covers names in Chinese and English, scientific names, morphology, uses and distribution. The list of wild grasses given in Table 22 will be seen to contain grass genera and species adapted to a wide range of ecoclimates, including the monsoonal. The remarks under the heading of 'uses' would seem to indicate that many of the grasses are regarded as of good quality for what must be rather a low level of animal husbandry and nutrition.

TABLE 22

Distribution of the wild grasses of Mainland China and adjacent territories[38]

Achnatherum sibiricum	N.W. and N. China, Manchuria, Inner Mongolia
A. splendens	N.W. China, Inner Mongolia
Agropyron cristatum	Sinkiang, Kansu, Tsinghai, Inner Mongolia, Shansi, Hopei, Manchuria
Agrostis alba	N. China, Inner Mongolia, S.W. China and Yangtze Valley
A. stolonifera	Kansu, Hopei, Honan, Chekiang, Kiangsi
Alopecurus aequalis	N. and S. China
Andropogon chinensis	Kwangtung, Yunnan
Aneurolepidium chinense	Manchuria, Inner Mongolia, Hopei, Shansi, Shensi, Sinkiang
A. dasystachys	Manchuria, Inner Mongolia, Hopei, Shansi, Shensi, Sikiang, Tsinghai, Kansu
Aristida adscensionis	N.W. China, Manchuria, Inner Mongolia
Arthraxon hispidus	Widespread

MAINLAND CHINA

Arundinella anomala	All areas except Sinkiang, Tibet, Tsinghai and Kansu
Bothriochloa ischaemum	Widespread
Bromus japonicus	Watersheds of Yangtze and Yellow Rivers
Calamagrostis epigejos	Widespread
Capillipedium parviflorum	S.W., Central and E. China
Chrysopogon aciculatus	Hainan Island, Kwangtung, Kwangsi, Yunnan, Taiwan
Cleistogenes chinensis	Inner Mongolia, N. and N.W. China
C. squarrosa	Inner Mongolia, N. and N.W. China
Clynelymus dahuricus	Manchuria, Inner Mongolia, Hopei, Honan, Shansi, Shensi, Tsinghai, Sikiang, Szechwan
C. nutans	Hopei, Shensi, Kansu, Tsinghai, Szechwan
Cynodon dactylon	South of Yellow River
Dactyloctenium aegyptium	South China
Deschampsia caespitosa	S.W., N.W., N. and N.E. China
Deyeuxia hakonensis	Chekiang, Anhwei
Digitaria ischaemum	All provinces in N. and S. China
D. sanguinalis	N. and S. China
Echinochloa crusgalli	Widespread
Eleusine indica	Widespread N. and S. China
Elytrigia repens	Sinkiang, Tsinghai
Eragrostis ferruginea	N. and S. China
E. pilosa	Temperate areas throughout China
E. poaeoides	Temperate areas throughout China
Eriochloa procera	Kwangtung, Hainan, Taiwan
Festuca arundinacea	Sinkiang—cultivated elsewhere
F. ovina	Uplands and prairies of Sinkiang
F. parvigluma	Shensi to the Yangtze Valley and S. China
F. rubra	Cold alpine areas in Manchuria, N., S.W., N.W. and Central China
Glyceria acutiflora	Kiangsu, Anhwei, Chekiang
Helictotrichon schellianum	N. China, Manchuria, Kansu, W. Szechwan
Hordeum brevisubulatum	Manchuria, Inner Mongolia, N. China, Sinkiang
H. violaceum	Kansu, Shensi, Tsinghai
Hymenachne amplexicaulis	Hainan, Yunnan
Imperata cylindrica	Widespread
Isachne globosa	Widespread except Manchuria
Koeleria cristata	Manchuria, N.W., N. and E. China
Leptochloa chinensis	E., Central and S. China and Szechwan, Kweichow and Shensi
Melica onoei	E. and N. China, Yunnan, Shensi
Miscanthus floridulus	S. China, Anhwei, Kiangsu
Panicum bisulcatum	Kwangtung, Kwangsi, Kiangsu, Chekiang, Yunnan, Kweichow, Szechwan, Hupei, Chilin, Liaoning, Heilungkiang

P. repens	Kwangtung, Hainan, Fukien, Taiwan, Chekiang
Paspalum distichum	Kwangtung, Kwangsi, Taiwan, Yunnan, Hupei, Kiangsu
P. orbiculare	Chekiang, Fukien, Kwangtung, Kwangsi, Kweichow, Yunnan
P. thunbergii	All provinces south of Yangtze
Pennisetum alopecuroides	Widespread
P. flaccidum	Manchuria, Inner Mongolia, N., N.W. and S.W. China
Phalaris arundinacea	Central and N. China, Manchuria, Inner Mongolia
Phleum alpinum	Manchuria, Szechwan, Tibet, Kansu, Shensi, Taiwan
Ph. paniculatum	Yangtze Valley and Shensi
Phragmites communis	Widespread
Poa acroleuca	E., Central and S.W. China, Kwangsi and Hopei
P. angustifolia	Yellow River watershed, Manchuria
P. annua	Most provinces
P. compressa	Hopei, Shantung, Kiangsi
P. nemoralis	Manchuria, Inner Mongolia, N.W. China, Hupei
P. pratensis	Manchuria, Inner Mongolia, Hopei, Shansi, Shantung, Kiangsi, Szechwan, Kansu
P. prolixor	Kiangsu, Chekiang, Hupei, Szechwan
P. sibirica	N.E., N. and N.W. China
Puccinellia tenuiflora	Manchuria, N. China
Roegneria kamoji	Widespread except Sinkiang, Tsinghai and Tibet
R. sinica	Shansi, Kansu, Tsinghai
R. turczaninovii	Manchuria, Hopei, Shensi
Rottboellia exaltata	Yunnan, Szechwan, Kwangtung, Kwangsi, Fukien, Taiwan
Sacciolepis indica	S. and S.W. China, Kiangsu, Chekiang
Schizachyrium brevifolium	S. Manchuria to Kwangtung and Hainan
Setaria lutescens	Widespread
S. viridis	Widespread
Spodiopogon sibiricus	N.E., N., E. and N.W. China
Sporobolus elongatus	S. and S.W. China, Kiangsu, Chekiang
Themeda triandra var. *japonica*	Widespread except Sinkiang, Tibet, Tsinghai, Kansu and Inner Mongolia
Trisetum bifidum	E. and Central China, Szechwan, Kwangtung, etc.
T. sibiricum	Manchuria, N. China, Kansu

Chapter VIII

FAR EAST U.S.S.R.

It will be seen that Figs. 4 and 5 show that the northern limit of the monsoonal ecoclimate in eastern Asia passes through Manchuria and even into the Far Eastern provinces of the Soviet Union (Ussuri and Zee-Buryat). Here, according to the Agricultural Atlas of U.S.S.R.,[224] the average temperatures for July are +15° to +22° C, but for January they are —7° to —20° C. The snow cover is 20–40 cm. deep. The precipitation of the warm period of the year is in some parts four times that of the cold period, in other parts two to four times. Rice is nevertheless grown in certain favoured localities in a region which is hardly one's normal conception of a monsoonal climate, even in the summer.

The study by V. N. Vasiljev[233] on the origin of the flora and vegetation of the Far East of U.S.S.R. and of eastern Siberia has not been available for study. A list of the grasses of Far East U.S.S.R. is given in Table 23.

TABLE 23

Grass genera and species of Far East U.S.S.R.[239]

Arthraxon hispidus	*Eriochloa villosa*
A. langsdorfii	*Microstegium nodosum*
Arundinella anomala	*Miscanthus purpurascens*
A. hirta	*M. saccharoides*
Asperella komarovii	*M. sinensis*
Cleistogenes (Diplachne) chinensis	*Muhlenbergia curviaristata*
C. hancei	*M. japonica*
Coleanthus subtilis	*Panicum bisulcatum*
Diarrhena japonica	*Rottboellia sibirica*
D. mandschurica	*Setaria gigantea*
Digitaria asiatica	*Spodiopogon sibiricus*
D. ischaemum	*Tripogon chinensis*
Dimeria neglecta	*Zizania latifolia*
Echinochloa caudata	*Zoysia japonica*
E. oryzicola	

Chapter IX

CHINA (TAIWAN)

Of the total land area of the main island of Taiwan (3·5 million hectares), 300,000 hectares are classified as grassland. M. Y. Nuttonson[159] gives figures for the altitudinal/climatic forest zones, indicating thereby the overall ecoclimatic conditions (Table 24).

TABLE 24

Taiwan. Altitudinal/climatic forest zones[159]

Type	Area in percentage of total area	Altitude in north m.	Altitude in south m.
Tropical	56	0–300	0–600
Subtropical	31	300–1,500	600–2,000
Temperate	11	1,500–2,700	2,000–3,500
Cold temperate	2	2,700–4,000	3,500–4,000

There are about 2 million hectares of forest land of which 10,000 hectares are covered with grass in the high mountains 2,000 m. above sea level. In the Ta-Hsüeh-Shan forest district in particular, the high mountain meadows cover 3,000 hectares in a total acreage of 60,000 hectares. Tsing Liu[223] has made an ecological study of a high mountain meadow on Mt. Hsiao-Hsüeh, as an initial study of the formation, composition and trend of succession in this type of grassland in Taiwan. These meadows are scattered throughout the forest region, but they are not to be regarded as a climatic climax.

Mt. Hsiao-Hsüeh is located in the central part of Taiwan 2,996 m. above sea level. The meadow studied, of about 250 ha., is on the southern slope at 2,500 m. altitude, surrounded by coniferous forests. The rocks are slate, quartzite and shale; soil grey-brown podzolic with gravel clay texture. Annual mean temperature is 9°–11° C, maximum 25·4° C, minimum —3·7° C. Winter is 4 months, and not very cold. Annual rainfall is 3,400 mm., mean relative humidity 81 per cent.

From 2,300 m. to 2,800 m. the dominant species are *Miscanthus transmorrisonensis* and *Pteridium aquilinum*. From 2,800 m. to 2,900 m. the dominant species are *Indocalamus niitakayamensis* and *Miscanthus transmorrisonensis*. Associated grasses under various micro-environments are *Agropyron formosanum, Arundinella setosa, Bromus morrisonensis, Deschampsia flexuosa, D. kawakamii, Festuca ovina, F. parvigluma* and *Brachypodium kawakamii*. The companion species are *Gentiana formosana, G. arisanensis, Solidago virga-aurea, Aletris foliata* var. *glabra, Anaphalis morrisonicola, Cirsium kawakamii, Swartia randaiensis* etc. Some small swampy areas are covered with *Juncus effusus, Brachypodium kawakamii, Agrostis morrisonensis, Deschampsia caespitosa* and *Carex kawakamii*. Shrubs are *Juniperus formosana, Stranvaesia niitakayamensis, Berberis kawakamii, Rosa morrisonensis, Gaultheria borneensis, Polygala japonica, Hypericum nagasawai, Vaccinium merrillianum* and *Rubus calycinoides*. Trees are *Pinus taiwanensis, Pinus armandii, Tsuga chinensis* and *Abies kawakamii* (above 2,800 m.). Most of the tree seedlings are pines. From circumstantial evidence, the next vegetal community will be pine forest.

According to statistical results of studies of the composition in plots nos. 1–10, *Indocalamus niitakayamensis* shows the highest degree of abundance of 62 per cent and constancy degree of class V, *Miscanthus transmorrisonensis* shows the next degree of abundance of 38 per cent and constancy degree of class IV. This community is a representative of the *Miscanthus transmorrisonensis/Pteridium aquilinum* Associes. In all 40 plots, *Miscanthus transmorrisonensis* shows the highest degree of abundance of 66 per cent, and constancy degree of class V, *Pteridium aquilinum* the next with abundance of 20 per cent and constancy degree of class IV.

The present meadow has been formed as a result of the burning of the original coniferous forest and its replacement by secondary pyrophytes such as *Indocalamus niitakayamensis, Miscanthus transmorrisonensis* and *Pteridium aquilinum*. With protection from fire, progression is characterized by the rapid in-

FIGURE 39

Ecological succession at high altitudes in Taiwan[223] according to Tsing-Liu

Succession A, above 2,800 m.:

ORIGINAL FOREST
Tsuga chinensis
Abies kawakamii
Association

Burn

PYRIC SUBCLIMAX VEGETATION
Indocalamus niitakayamensis
Miscanthus transmorrisonensis
Associes

Burn

Indocalamus niitakayamensis
Miscanthus transmorrisonensis
Associes

CLIMAX FOREST
Tsuga chinensis
Abies kawakamii
Association

Succession B, below 2,800 m.:

ORIGINAL FOREST
Tsuga chinensis
Chamaecyparis taiwanensis
Association

Burn

Miscanthus transmorrisonensis
Indocalamus niitakayamensis
Associes

Competition

PYRIC SUBCLIMAX VEGETATION
Miscanthus transmorrisonensis
Pteridium aquilinum
Associes

Burn

Miscanthus transmorrisonensis
Pteridium aquilinum
Associes

Burn

Pinus taiwanensis
P. armandii
Associes

CLIMAX FOREST
Tsuga chinensis
Chamaecyparis taiwanensis
Association

vasion of pines (*Pinus taiwanensis, P. armandii*), hemlock (*Tsuga chinensis*) and *Abies kawakamii* above 2,800 m. However, the pioneering pines do not form much of a canopy, because the matted grasses hinder the germination of the few tree seeds that are available. Burning frequently recurs and the *Miscanthus/Pteridium* community will be preserved.

Tsing Liu provides a diagram of this succession in these mountain meadows below and above 2,800 m. altitude (Fig. 39).

The main species in the grasslands on lower slopes is *Miscanthus sinensis*, probably with a group of consocies. In drier areas on lower or higher slopes the main species is *Imperata cylindrica* var. *koenigii*, with a group of consocies.

Chapter X

JAPAN

Brief reference is made to the secondary types of grass cover, because, as we have already seen, Japan comes on the northern limit of monsoon Asia with its humid tropical Pacific summers alternating with cold to sub-arctic Siberian winters (in Hokkaido, along the coast of the Sea of Japan, and in the mountains generally). A summary of Japanese literature up to 1955[248] has shown that the main types of grass cover in a forest climax are dominated by *Miscanthus japonicus*, *Sasa* or the dwarf bamboo, and *Zoysia* on poor sites at high altitudes. Subsequent literature will be found in the *Journal of the Japan Grassland Society*. The reclamation of these economically unproductive types of grass cover to pastures composed of superior European species of grasses and clovers has been described elsewhere.[250]

Chapter XI

MALESIA*

It is preferable to use the Latin term, Malesia, to avoid confusion with the political entity, Malaysia, and to include, following C. G. G. J. van Steenis, within this term all these parts of Western and Central Malesia, defined below, which are situated between south-eastern continental Asia on the one side, and Papua and New Guinea, New Britain and Northern Australia and Northern Queensland on the other.

South-eastern continental Asia, including Burma, Thailand, Indo-China, South China and Hainan, is an area of which the major surface is subject to annual drought, certain ever-wet foci excepted; southwards this climatic regime is found to the south of Peninsular Thailand about the isthmus of Kra.

The western part of Malesia, including Malaysia, Sumatra, Borneo and West Java, forms a colossal continuous area with an ever-wet climate; as far as may be concluded from the fossil record this has been so at least onwards of the Pliocene, and probably even the Miocene. In it are only a negligible number of very local dry spots caused by rain-shadows cast by mountains (Pleihari in South-east Borneo, Sumatra West Coast, Tapanuli, Atchin).

The central part of Malesia, including the Philippines, Celebes, the Moluccas, Central and East Java, and the Lesser Sunda Islands, is an area characterized by a climatic mosaic of ever-wet and seasonal. The amount of seasonal area differs from island to island. In the Philippines the western faces of almost all the western islands are subject to a dry monsoon. In Celebes the largest areas with a dry season are found in the southern extremities of the two southern peninsulas, but in Central and North Celebes there are only very local areas which, through

* This section is based almost entirely upon literature and personal communications from Prof. C. G. G. J. van Steenis and Charles Monod de Froideville (1964).

topographic causes, have a dry season. The Moluccas show a mosaic which is predominantly ever-wet, but with isolated spots or local areas which are seasonal; they are found in Halmahera, Buru, Ceram, the Aru Islands, etc. Central and East Java, however, possess more seasonal area than ever-wet and in East Java the ever-wet spots are simply caused by local topography, the southern sides of the volcanoes having an ever-wet climate due to the rain given off by the monsoon winds ascending the slopes. The Lesser Sunda Islands have a similar mosaic in East Java.

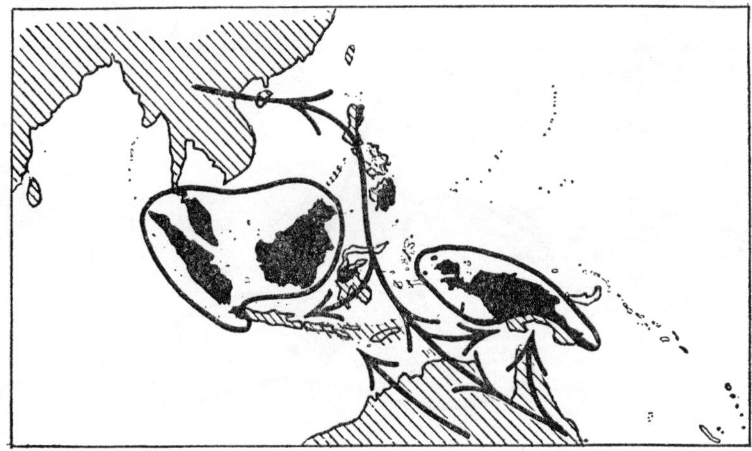

FIGURE 40

The two big cores of ever-wet rain-forest on the Sunda Shelf and New Guinea (black), barriers for distribution of seasonal plants: the hatched areas either dry or consisting of a mosaic of ever-wet and dry areas and spots[228]

New Guinea is almost entirely ever-wet, the southern coastal part from Frederik Hendrik Island to Moresby excepted; besides, there are along the coasts some local dry pockets caused by rain-shadow. New Britain is also ever-wet in general, but shows in various places dry pockets. North Australia and North Queensland are also largely subject to a seasonal climate.

Types of Grassland

In the vegetation map of Malesia published in 1958,[232] three of the types are related to grassland, namely:

Type 11: Savannahs, ranging from grassland with scattered trees to savannah woodland. They are generally seral types due to fire in seasonal climatic conditions. The dominant species include: *Melaleuca* (eastern Malesia), *Eucalyptus* and *Acacia* species (Flores, Timor, Alor, Wetar, south New Guinea), and species of the palm genera, *Borassus* and *Corypha* (east Java, Lesser Sunda Islands).

Type 12: Grassland, of various species mixed with herbs and ferns; geotypes abundant; presence invariably due to fire and

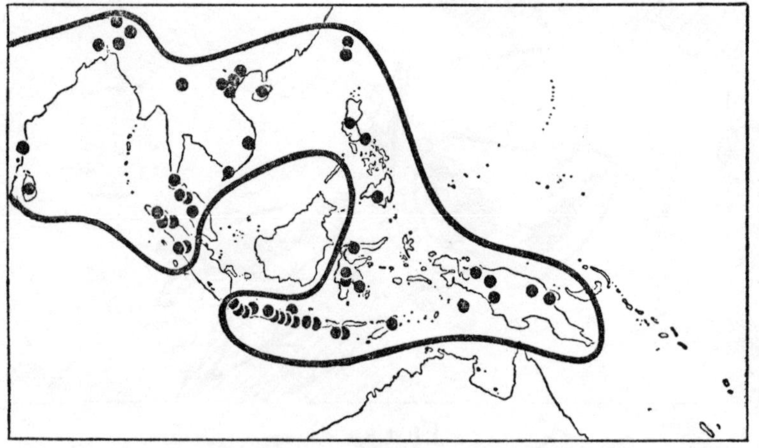

FIGURE 41

Smithia sensitiva, a species with preference for a feeble dry season as occurs in N. Sumatra, N. Malaysia and W. Java, and scattered localities in New Guinea[228]

absence of shade, especially abundant where climate is seasonal; indifferent to altitude. This mixed grassland is often called lalang, alang or cogon; these names refer strictly to species of *Imperata*, which require rather good non-inundated soil in a rather wet climate with frequent fires. Most lalang grasslands contain little *Imperata*.

Type 13: Alpine grassland, with herbs and prostrate dwarf shrubs above the climatic timberline, in primary condition found only in the highlands of the Main Range of New

Guinea, now extending far below the original limit because of hunters' fires.

The 'anthropogenic' grasslands of Papua and New Guinea (Chapter XII) cannot possibly be considered a natural climax, but are maintained as grassland as a result of the agricultural and burning activities of the native population.[189] The grassland may have somewhat of a savannah status due to the occurrence of fire-tolerant trees 9–12 m. high, including *Antidesma ghaesembilla, Albizzia procera, Cordia dichotoma, Nauclea orientalis*

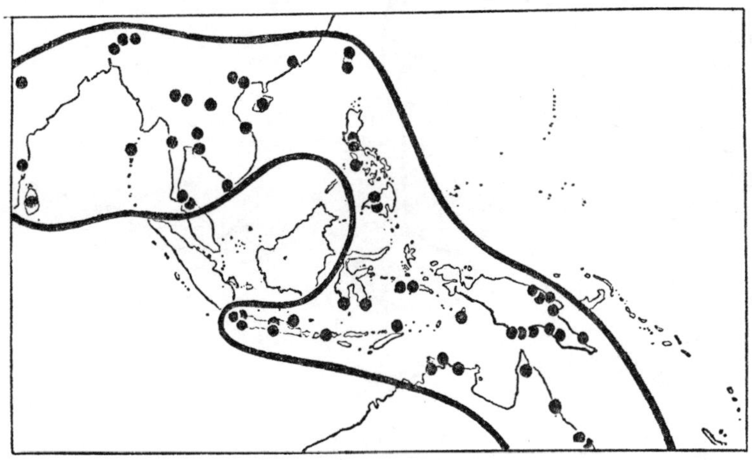

FIGURE 42

Pycnospora lutescens, a species requiring a pronounced dry season; absent from the Sunda Shelf and only in strictly seasonal spots in Papua[228]

and *Cycas media*. Floristically and structurally these grasslands fall into either a tall or short category. Tall grass communities are natural regrowth phases following closely upon forest clearing. In North Papua an association of *Saccharum spontaneum* with *Imperata cylindrica* is frequent, in Ramu Valley *Ophiuros exaltatus*/*Imperata cylindrica*, mixed canegrass communities up to 3 m. high in the populated Mafrik area of Sepik District (*Polytoca macrophylla, Sorghum halepense, Saccharum spontaneum, Coelorhachis rottboellioides, Pennisetum macrostachyum* and *Coix lacryma-jobi*).

Large areas in the highlands (particularly Western Highlands) are at present covered with a 3·5m. sword grass; this may pass through a lengthy series of floristic degradations over perhaps 100 years or more until short grasslands are established after a long or intense history of interference. Its development is more rapid in areas of unfavourable habitat, but everywhere it represents a disclimax—an induced and stabilized community.[188]

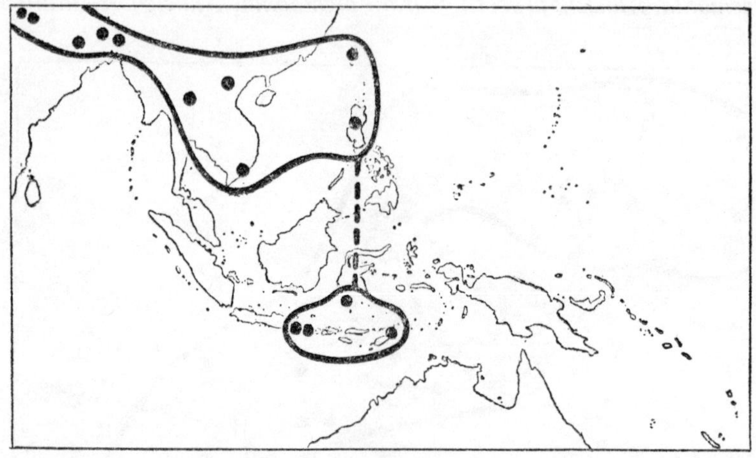

FIGURE 43

Smithia ciliata, a montane species requiring a rather marked dry period; between the Philippines and Java there is still a locality in Celebes. The area is semi-disjunct[228]

Distribution of species

Van Steenis finds that, in mapping the areas of distribution of the species of the seasonal flora, it appears that they follow the areas and spots with seasonal climates. They occur in a large number of localities in continental south-east Asia, in isolated dry spots in central Malesia (therefore most common in Java and the Lesser Sunda Islands, and more frequently again in south New Guinea, the Bismarcks and Australia). The pathway of distribution of drought plants through Malesia is broken up into a sequence of smaller or larger stepping-stones; apart from the insular character of Central Malesia, it is only small parts of these islands which offer a suitable ecoclimate.

Studies made of the sub-family Papilionaceae for the Flora Malesiana[228] have shown that there is one group of the sub-family which consists of genera which have a worldwide tropical distribution—even many of their species show a very wide range

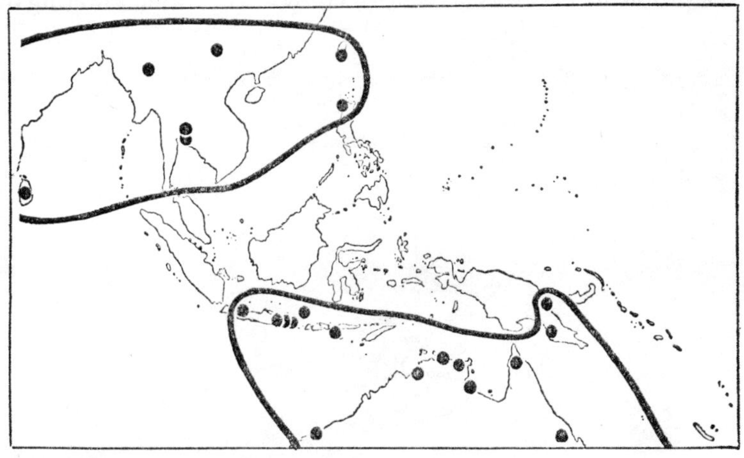

FIGURE 44

Rhynchosia minima, a species requiring a strong dry season, with an obvious disjunction[228]

of distribution. In general, these species do not belong to the rain forest, but to areas subject to an annual dry season of varying duration. It thus becomes a question how these legumes, and by the same token the grasses of the Malesian region, that are adapted to seasonal variations in climate could cross what van Steenis calls the ever-wet tropics between continental Asia and Australia (Figs. 40–45).

How could these genera and species of legumes and grasses become dispersed over these now remote stepping-stones, which are separated by distances too wide for bridging by natural means of dispersal? Dispersal by man has to be considered, but these plants have in many cases occupied such vast areas already one or two centuries ago that this proposition seems unlikely.

This same drought pathway has been occupied not only by species which come from Asia, but also by those from Australia, and by those native to Malesia.

The solution must be found either in a drier climate or more stepping-stones in the past to enable the active dispersal of species which would now be impossible. Van Steenis brings forward the influence of the Pleistocene Ice Age as the solution of the problem. This was accompanied by a worldwide lowering of

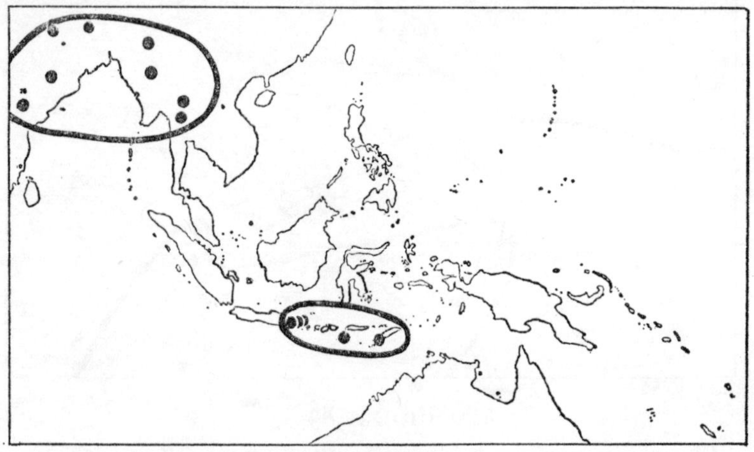

FIGURE 45

Rhynchosia rothii, a species bound to a severe dry season, in Malaysia restricted to the driest area. The disjunction is still wider than that shown in Figure 44[228]

the level of the sea, which resulted in the merging of the land masses of West Malesia on the Sunda Shelf into one very large peninsula, and the merging of Northern Australia with the Aru Islands and New Guinea in the Papuasian Sahul Shelf area. The increase of the land masses changed the wind regime and air humidity in Malesia. The result would be that there must have been distinctly more and larger areas of periodical drought between the two larger rain-forest areas in West and East Malesia respectively. These two ever-wet cores remained stable, but locally and along their margins there would be more local dry spots than there are at present, now faintly shown by some indicator plants tolerant of a feeble dry season.

With the post-glacial rise of the sea level, the situation as it is at present gradually returned. Many stepping-stones were overwhelmed by rain forest, but some remained intact, particularly on the rain-shadow slopes and valleys. The wider spacing of the stepping-stones was caused partly by the post-glacial increase of humidity, but also by the submersion of large areas, notably in the South China Sea, which caused a large gap to arise between the drought areas of Viet Nam and the Philippines.

A (geologically) rather recent exchange of drought plants between Asia and Australia fits well, according to van Steenis, with

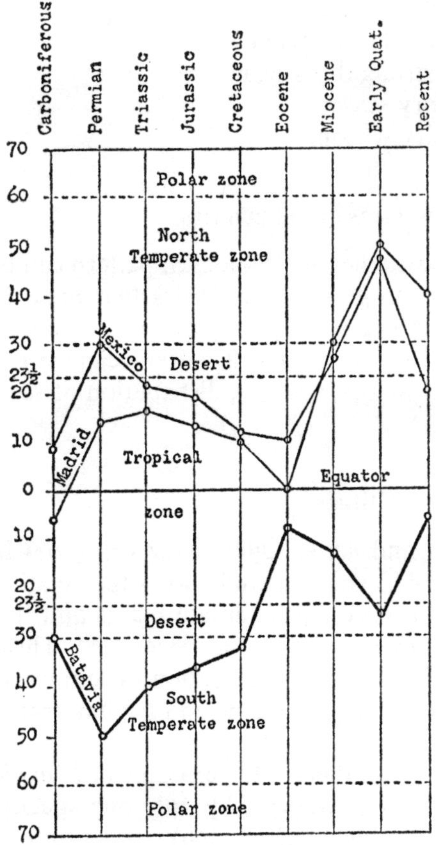

FIGURE 46

Latitudinal variations of the Malesian Archipelago from the Carboniferous to the present, from C. G. G. J. van Steenis[230] after Köppen and Wegener, *Die Klimate der geologischen Vorzeit* (1924)

the fact that practically all these species are exactly the same as are found on the continents, thus indicating recent dispersal. A similar observation was made by Warburg[242] on the savannah plants in the dry areas of south New Guinea which are specifically not distinct from those of Australia and must be of 'recent' arrival.

Species may be sorted into drought classes, as follows, and certain species may be used as distinct indicators for these classes. Some species needing only a feeble dry period are found very locally in some isolated dry pockets in the rain-forest area, but those needing a strong or long dry period are rarest and show the most disjunct distribution:

Indifferent to climate
Feeble dry season
Pronounced dry season
Rather strong dry season
Strong dry season
Severe dry season.

Ecology of Grass Communities

In reply to the question whether any ecological pattern can be discerned with regard to the grass communities that arise or occur as subclimaxes in areas with sparse or no tree cover, Monod de Froideville (1964) states that a pattern can be observed in relation to three factors, altitude, distribution of rainfall and soil.

Altitude

The location of extinct and active Quaternary volcanoes in Malesia is shown in Fig. 47, while Fig. 48 shows the relation between the distribution of Malesian mountain plants and this Quaternary volcanism (from C. G. G. J. van Steenis[230]). These studies indicate the principal tracks of migration of temperate genera into Malesia by the Sumatran, Luzon and Papuan tracks (Fig. 49).

Before turning to the grasses, there is the example of *Primula* as a dicotyledon, represented in Malesia by only one species, *Primula prolifera* Wall (= *P. imperialis* Jungh). Its altitudinal

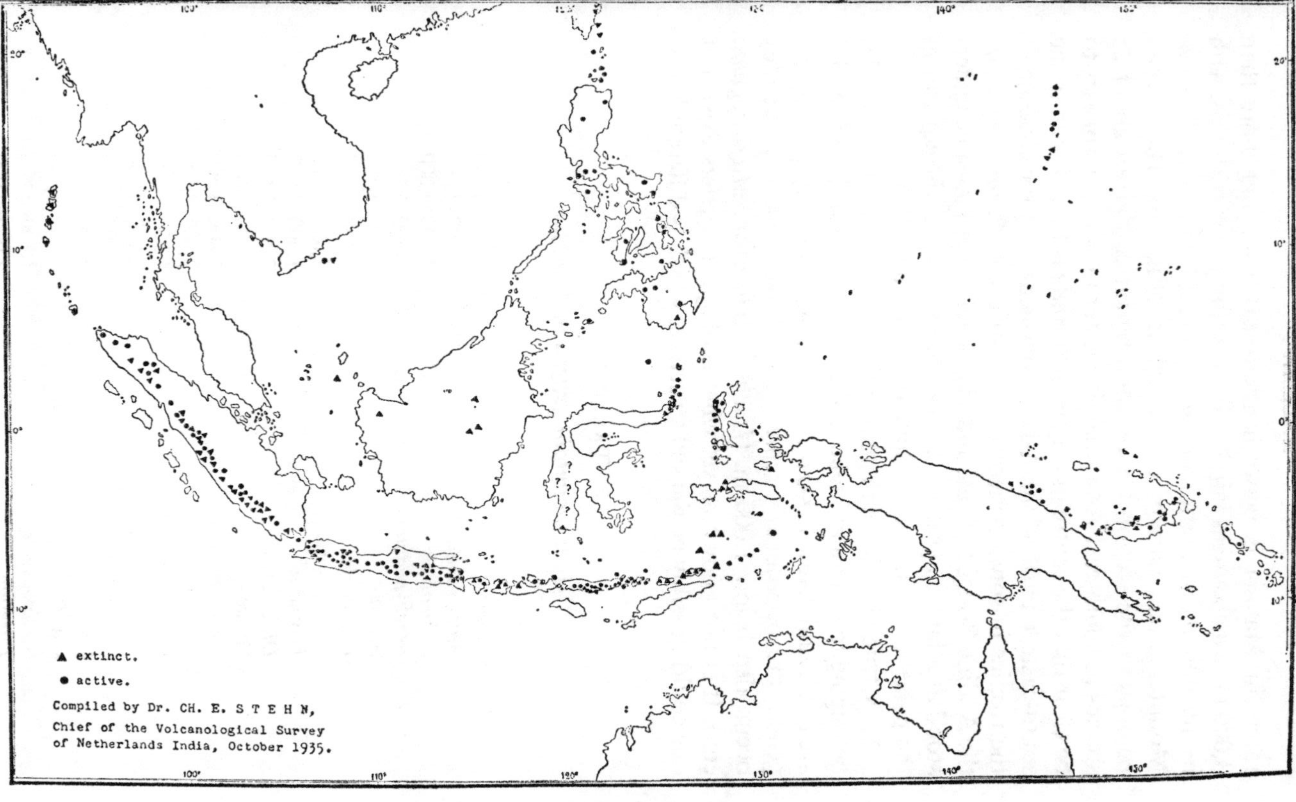

FIGURE 47
Location of Quaternary volcanoes in the Malesian region[230]

range in Malesia is between about 2,050 m. and more than 3,000 m. which agrees fairly well, according to van Steenis, with its altitudinal distribution in Asia, where it occurs in the Khasya Mountains and part of the Himalaya (Fig. 50). This species belongs to the section Candelabra, which comprises about 25 species in south-east Asia, mainly western China. It appears to be clear that the centre of origin of *Primula* is to be found in the south-eastern parts of the Asiatic continent, that the crossing of the tropics in south America (by a variety of *P. farinosa* along the Andine Bridge to Chile and Tierra del Fuego) must be understood as a later migration, as also the crossing of the equator in the palaeotropics, that is, Sumatra (van Steenis[229]).

It is true that in high mountain areas there occur species belonging to genera of higher latitudes (Table 25). These species are not found below 1,000 m. On the other hand, typical tropical genera are occasionally represented above 2,000 m., but they occur also below 1,000 m. (*Isachne*, *Arthraxon*, *Microstegium*, etc.). It is not known as yet what ecological barriers exist, but one could possibly be the soil structure, which is different below

TABLE 25

High mountain genera in Malesia

	Metres
Agrostis	1,300–4,000
Anthoxanthum	2,000–3,500
Aulacolepis	2,400–3,000
Brachypodium	1,700–3,600
Bromus	1,700–3,560
Danthonia	2,400–3,800
Deschampsia	2,000–4,000
Deyeuxia	1,500–4,000
Dichelachne	2,400–3,680
Festuca	1,600–4,050
Helictotrichon	1,800–3,500
Hierochloe	2,100–3,700
Microlaena	1,500–3,700
Monostachya	2,700–4,100
Muhlenbergia	1,300–2,400
Poa	2,000–5,000
N.B. *Poa annua* (introduced)	900–1,300
Streblochaete	1,800–2,800
Trisetum	2,000–4,000

and above a certain altitude (approximately 1,800 m.), as demonstrated by the Pedological Institute at Bogor. Above the critical line the soil tends to form peat.

In the mountain forest between 2,000 m. and 3,000 m. altitude the normal physiognomy is a stunted, very mossy forest, dripping wet because it is in the cloud zone. It would be unimaginable to burn it, but in exceptionally dry years the moss is dry and extremely inflammable, and has been observed burning. Forest regeneration is slow because of the slow growth due to high altitude. There is thus always a fringe of dead, charred timber. In the next exceptionally dry year this charred fringe will feed a new forest fire, and the grassy glades will thus expand.

It is striking that a number of temperate genera do not occur in Malesia, such as: *Agropyron*, *Aira*, *Alopecurus*, *Avena*, *Catabrosa*, *Cynosurus*, *Glyceria*, *Hordeum*, *Koeleria*, *Melica*, *Oryzopsis*, *Polypogon*, *Stipa*, etc. The genera of wet habitats generally have a wide distribution, e.g. *Phragmites*, but even these (*Catabrosa*, *Glyceria*) are not all represented in Malesia.

Nearly all the high mountain genera mentioned here are of the festucoid type, confined to temperate regions. The genera *Monostachya* and *Danthonia* are of a different type, but they belong to the tribe Danthonieae which is also confined to temperate regions.

Distribution of rainfall

It is noticed that in Java *Heteropogon contortus* occurs only in places with a pronounced dry season (up to 10 days with rain in the 4 driest months). In this connection, *Heteropogon contortus* contrasts with *Imperata cylindrica*, which grows in regions without a pronounced or prolonged dry season. Perhaps their areas coincide with those of deciduous tropical monsoon forest and of evergreen tropical rain forest, respectively.

Heteropogon contortus is not the only species confined to the monsoon climate. There are a number of other members of the tribe Andropogoneae with the same restricted distribution, including *Heteropogon triticeus*, *Dichanthium* spp., *Sehima nervosum*, *Bothriochloa* spp., *Eulalia* spp., *Themeda australis*. On the other hand, many grasses occur in both climates, e.g. the giant *Themeda* spp. and *Th. arguens*.

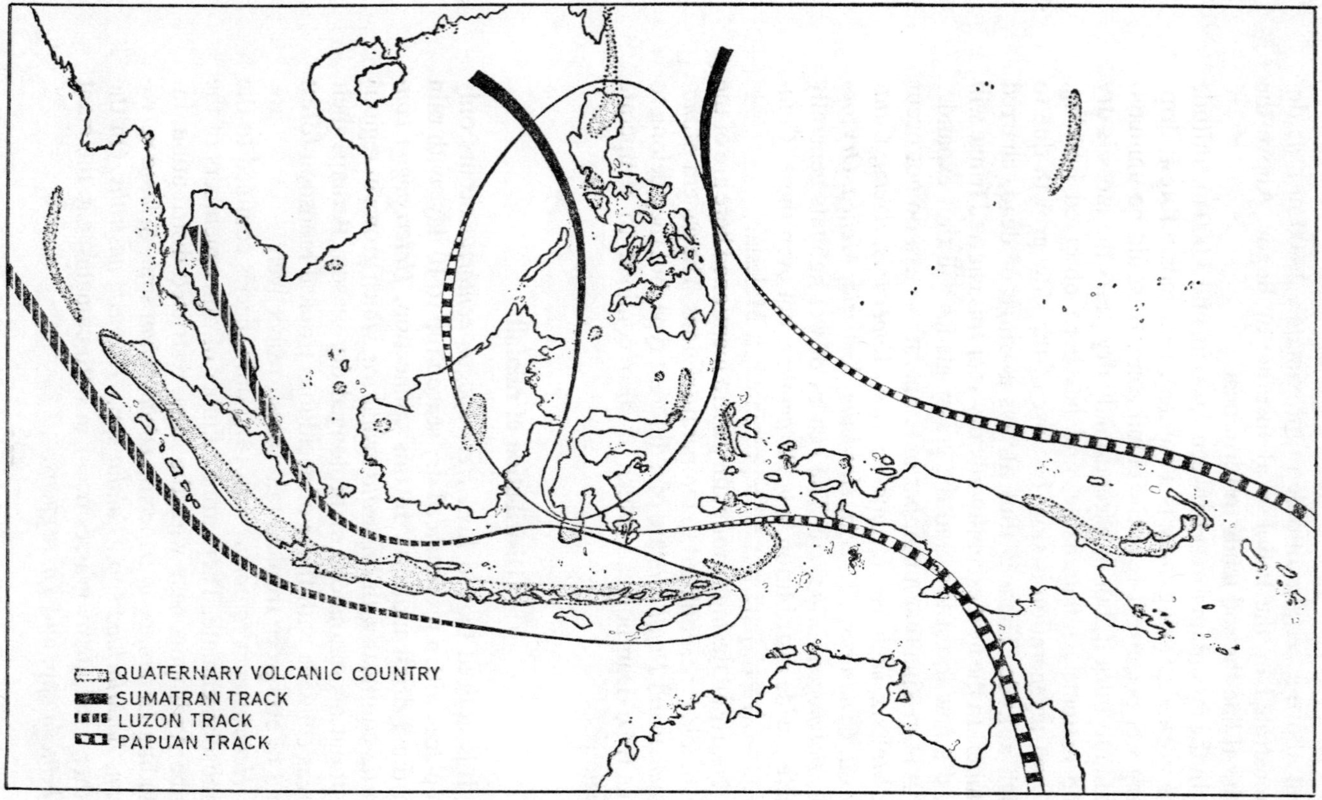

QUATERNARY VOLCANIC COUNTRY
SUMATRAN TRACK
LUZON TRACK
PAPUAN TRACK

FIGURE 48

Saccharum spontaneum should not be mentioned here, because its distribution is not so much restricted by climate as by the need for a constant soil water level. It occurs throughout Malesia.

The influence of exceptionally dry years, which occur usually once in 5 and certainly once in 10–20 years, may be considerable because in such years the Malesian area is subject to extensive burning. Even boggy Bornean peat forest is then dry and inflammable. Extensive shallow lakes in East Borneo are now covered with the so-called 'kumpai' grassland (drifting sods); they were once swamp forests of *Shorea belangeran* and other swamp forest trees. Fishermen fire this grassland in the exceptionally dry years to convert the shady forest without fish into open lakes with fish.

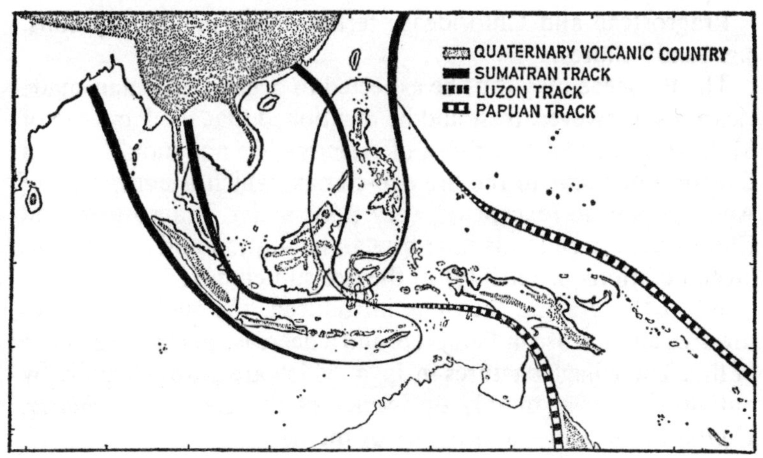

FIGURE 49

Principal tracks of migration of temperate genera into Malesia. The dotted area is the zone above 1,000 m. altitude. The thickness of the lines indicates to some degree the number of genera belonging to the respective tracks[229]

Soil

A number of grasses grow in saline habitats, such as the sea shore and inland salt marshes. Some species are obligatory halophytes (*Xerochloa imberbis*, *Paspalum vaginatum*, *P. distichum*, *Diplachne fusca*, *Sporobolus virginicus*, *Thuarea involuta*), others, like *Ischaemum muticum* and *Zoysia matrella*, occur also in non-saline places. *Chloris barbata*, a common ruderal in coastal places, was also found far inland in Timor, in saline dried mud.

It is certain that only a limited number of the grasses of Malesia grow in extremely poor soil; they include species of the genera *Arundinella*, *Eriachne*, *Apocopis* and *Eremochloa*. The species belonging to the tribes Andropogoneae, Paniceae, Eragrosteae and Chlorideae, which form the bulk of the Malesian grasses, are as a rule more exacting, but each group in a different way, as follows:

Andropogoneae: well-aerated virgin soils, such as landslides.

Paniceae: fertile, moist soils, such as occur on river banks and in open spaces in forests.

Eragrosteae and Chlorideae: fertile, more or less dry soils, e.g. sandy shores.

The Paniceae may thus be expected to occupy new man-made clearings. However, it should be mentioned that the Paniceae of Malesia, in contrast to those of America, do not stand fire and thus do not occur in the fire sub-climax, which is composed of Andropogoneae (except on very poor soils). Eragrosteae and Chlorideae are ruderals near human habitation and along trails (accumulations of fertility) in the fire sub-climax.

When the grazing factor is introduced, the succession goes towards an increase in Paniceae, and a decrease in the occurrence of fire. The village pastures in Java, which are grazed heavily by buffaloes, consist mainly of species of the genera *Brachiaria*, *Digitaria*, *Paspalum* and others of the tribe Paniceae.

Fire

The monsoon forest is always thinner and lighter than the rain forest, with lower trees and hardly any secondary storeys beneath the canopy; grass is of course found on the ground, while it is absent in the primary rain forest. The conversion of the monsoon forest to grassland through fire is easy, and various types of savannah-like vegetation can originate: orchard savannah, savannah woodland, coulisse savannah, etc. In the Malesian monsoon forest burning is a universal practice, though it varies in intensity and frequency. Many woodland savannahs are characterized by fire-resistant trees (*Eucalyptus*, *Casuarina*, *Melaleuca*, etc.). If the intensity or frequency of fire is high, the ultimate fire climax is tall grassland. If burning is infrequent, such pioneer stands may develop into more or less monospecific stands. Monod de Froideville has provided particulars (Tables 26 and 27) of grass cover types investigated by him in Java, Timor and Sumba.

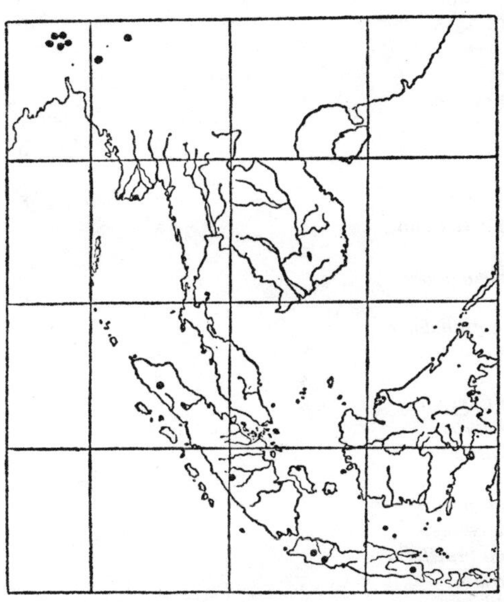

FIGURE 50

Distribution of *Primula prolifera* in Khasya Mountains, eastern Himalaya and Sumatra, at elevations between about 2,050 and more than 3,000 m.[229]

Parallelism between Forest and Grass

Until the types of climax formations have been described in the Flora Malesiana, it is difficult to say whether there is any parallelism between a particular forest type and an associated type of grass cover. Even after publication of the flora, it may be difficult to establish a parallelism, but already some vegetation

TABLE 26

Types of pasture subject to fire in regions with pronounced dry season (Monod de Froideville, 1964)

Species	I	II	III	IV
Chlorideae				
Cynodon dactylon		o		
Paniceae				
Digitaria agyrostachya	o	o	o	o
Paspalum cartilagineum			o	
Setaria pallide-fusca			o	
Andropogoneae				
Bothriochloa glabra	x		x	
B. pertusa	x	x		
Chrysopogon aciculatus		o	o	
Ch. tenuiculmis				a/p
Cymbopogon rectus			x	x
Dichanthium caricosum	x	x		
D. erectum	x			x
Eulalia leschenaultiana		x		x
E. trispicata			x	
Hackelochloa granularis			a	
Heteropogon contortus	x	x	x	x
H. triticeus	x	x	x	x
Hyparrhenia filipendula	x			
Imperata cylindrica			x	
Ischaemum barbatum			x	
I. timorense			o	
Iseilema minutiflora				a
Ophiuros tongcalingii			x	
Schizachyrium brevifolium		a	a	a
Sehima nervosum	x			x
Sorghum nitidum			x	
Themeda arguens	x			
Th. australis	x		x	

I. Pasture on tertiary limestone, altitude 800 m., annual rainfall 1,800 mm. Timor Island.
II. Pasture on light marl, altitude 400 m., annual rainfall 1,200 mm. Same island.
III. Pasture on hills of basic eruptiva and sediment, rich in lime, altitude 400 m., annual rainfall 1,800 mm. Occasionally cultivated. Sumba Island.
IV. Pasture on sediments, rich in lime, coastal area, annual rainfall 750 mm. Sumba Island.

x Pyrophytes.
o Elements of non-pyrogenous pasture.
a Annuals.
a/p Annual or perennial according to circumstances.

TABLE 27

Types of pasture not subject to fire (Monod de Froideville, 1964)

Species	I	II	III	IV	V	VI
Eragrosteae						
Eleusine indica		o	o			
Chlorideae						
Cynodon dactylon	o	o		o	o	o
Sporoboleae						
Sporobolus diander		o				
Paniceae						
Axonopus compressus	o					
Brachiaria sp.		o				
B. paspaloides				o		
B. reptans				o		o
Digitaria agyrostachya		o	o	o		o
D. fuscescens	o					o
D. longiflora	o					
D. pseudo-ischaemum				o		
Panicum repens	o	o				
Paspalum commersonii	o			o		
Sacciolepis indica	o					
Andropogoneae						
Bothriochloa glabra	x			x	x	
Chrysopogon aciculatus	o	o	o	o		o
Dichanthium caricosum		x		x	x	x
Dimeria ornithopoda	o		o			
Eulalia leschenaultiana						x
E. trispicata			x			
Imperata cylindrica	x					
Ischaemum timorense		o	o		o	
Polytrias praemorsa	o	o			o	
Themeda arguens	x			x	x	
Th. australis			x			

I. Pasture at Bogor (Java).
II. Sportsfield at Waikabubak (Sumba Island).
III. Village common at Anakalang (Sumba Island).
IV. Pasture at Loli (Timor Island).
V. Pasture at Noilmina (Timor Island).
VI. Pasture at Bidjeli (Timor Island).

I in region without pronounced dry season.
II–VI in region with pronounced dry season.
All at low altitude—400 m.

x Pyrophytes.
o Non-pyrophytes.

types can be seen in which grasses and trees are associated. These are all sub-climax communities in respect to both the trees and the grasses, in which fire is the stabilizing factor.

(1) *Pinus merkusii* vegetation in northern Sumatra. Grass undergrowth consists mainly of giant *Themeda* spp. and *Saccharum spontaneum*.

(2) Vegetation of the 'Padang Bolak' ('Big Plain') in central Sumatra, established in rain forest climate on old sterile and acid soils of sand and liparite sand. Average rainfall: 2,300 mm. per year, with 2 months of less than 100 mm. Dry wind from June to September. Trees: *Randia*, *Fagraea*, *Phyllanthus*, *Baeckea*. Grasses: *Eriachne triseta* and *Arundinella setosa*. No *Imperata* or *Themeda*. Low altitude.

(3) *Eucalyptus* vegetation in Timor. Prolonged dry season. Undergrowth of *Heteropogon contortus*. Low altitude.

(4) Restricted area of heavy soil in Timor, with *Acacia leucophloea*. Soil cover of *Sclerachne punctata*. Low altitude.

(5) *Casuarina junghuhniana* in Java is associated with the grasses *Streblochaete longiaristata*, *Festuca nubigena* and *Pennisetum alopecuroides*. Mountains.

Grazing factor

As regards succession in different types of grass covers under the influence of biotic or other factors, Monod de Froideville refers again to the general pattern:

In wetter climate the Andropogoneae consist of a virtually pure stand of *Imperata cylindrica*. Since the aerial parts of this species do not stand repeated trampling by livestock, it is suppressed by adequate stocking and the Paniceae can come in. But in drier climates the stocking results in the replacement of the original Andropogoneae (*Heteropogon contortus*, tall *Themeda* spp.) by short species of the same tribe, and only in the last instance do the Paniceae come in. In a sense the transition from fire sub-climax to short-grass range is more abrupt and possibly more difficult to establish with *Imperata* than with other vegetation. There are many patterns, according to different regions or even within the same region. Even in a field each species occupies its special habitat, so that each individual field shows a definite pattern.

FIGURE 51

Succession observed in the Humboldt Bay region of New Guinea according to M. Monod de Froideville

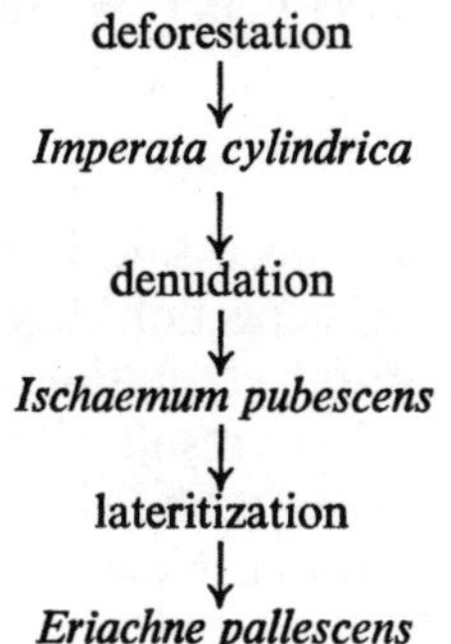

Certain limitations with regard to *Imperata cylindrica* are:

(*a*) In regions of prolonged dry season the species only occurs as a (persistent) weed in cultivation. Where the rotation includes pasture, an *Imperata* stand will cover the field soon after the maize or other cereal crop has been harvested, but will disappear again after being trampled by grazing livestock.

(*b*) The species does not stand flooding for a prolonged period.

(*c*) It does not thrive on stony soil.

(*d*) It does not stand trampling and regular cutting.

Chapter XII
PAPUA AND NEW GUINEA

In addition to what has been said in Chapter XI regarding Papua and New Guinea as part of Malesia, valuable information regarding the natural vegetation and grasslands of the territory has been obtained as a result of the work of the Division of Land Research, Commonwealth Scientific and Industrial Research Organization, Australia. New Guinea is one of the last great land masses to be explored biologically. According to R. G. Robbins[188] the vegetation formations range from coastal mangroves and strand forest through a range of humid lowland rain forests and swamps to mountain oaks and beeches and finally *Rhododendron* shrubberies and alpine meadows.

Due to favourable climate and soils, perhaps 80 per cent of the region is covered with the floristically rich and luxuriant forests. Only in a southern zone of lower rainfall near Port Moresby and Fly River does one find monsoon forest and dry communities of savannah woodland of scattered eucalypts with a ground cover of grasses, regarded as extensions of north Australian vegetation. Although the genera and species at higher elevations along the central mountains show affinities with those of Tasmania and New Zealand, most of the plants reflect an Asian origin, bringing New Guinea into the Indo-Malayan or Indo-Malesian region.

This becomes clear when one extracts information regarding grassland communities from the four Reports of the Division of Land Research that are available so far.[90, 190, 217, 218] The various types of savannah, mid-height grassland and tall grassland are not reliable indicators of climate and soil conditions because of the overriding influence of repeated burning. The areas covered are shown in Fig. 52.

Buna-Kokoda

Three different disclimax grassland alliances have been recognized by B. W. Taylor.[217] Scattered trees are found in all three types, the most common species being *Antidesma ghaesembilla*, *Nauclea orientalis* and *Timonius timon*. Over 100 other tree species have been seen in these grasslands but these are generally of rare occurrence.

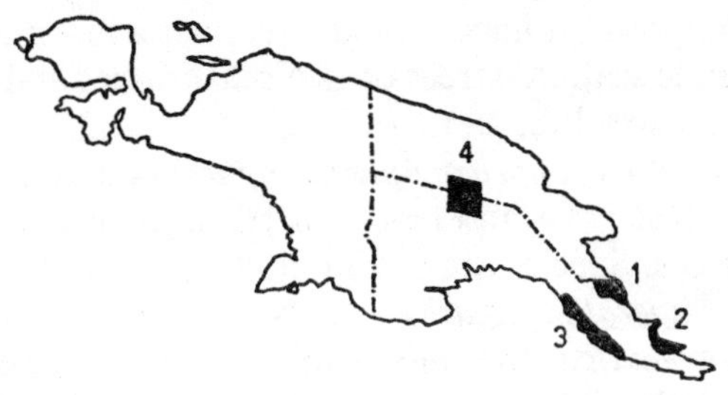

FIGURE 52

Location of areas surveyed by Division of Land Research, C.S.I.R.O., in Papua and New Guinea

1. Buna-Kokoda
2. Wanigela-Cape Vogel
3. Port Moresby-Kairuku
4. Wabag-Tari

(i) *Saccharum spontaneum/Imperata cylindrica* Alliance. After the regular yearly burning the dominant grass species *Saccharum spontaneum*, *Imperata cylindrica*, *Ophiuros exaltatus* and *Coelorhachis rottboellioides* rapidly regenerate. Any combination of these four species may be locally dominant but *Saccharum spontaneum* is the most abundant species. Other grass species common in this alliance are *Apluda mutica*, *Sorghum nitidum* and *Hymenachne amplexicaulis*. The occurrence of other herbaceous species is more variable and tends to be more frequent in drier areas where up to 30 species have been recorded from a small area. In a mature stand, which may exceed 3 m. in height before the annual firing, generally only a few species are present, including *Pueraria* spp., *Merremia hirta*, *Passiflora foetida* and a fern ('kukusa'). Common species in a recently burnt stand include *Pygmaeopremna sessilifolia* and *Uraria lagopodioides*. In addition to the three tree species common in grasslands, *Ficus calopilina* is also a common tree in this alliance.

(ii) *Themeda australis/Imperata cylindrica/Coelorhachis rottboellioides* Alliance. This alliance consists of a sequence of communities probably depending on the intensity of burning. It varies from a community dominated by *Imperata cylindrica* or by *Saccharum spontaneum* to a community dominated by *Themeda australis* and *Coelorhachis rottboellioides*. At this later stage the community rarely exceeds 1 m. in height, and the floristic composition may vary greatly over a few metres. A few stands include scattered trees of *Albizzia procera* and *Eucalyptus tereticornis*, exceeding 18 m. in height.

(iii) *Imperata cylindrica/Apluda mutica* Alliance. This community is dense grassland less than 1·5 m. high with occasional scattered trees. Species are few in number compared with other grassland. *Imperata cylindrica* may be dominant or co-dominant with *Apluda mutica*, the only other common species. The tree species include common grassland species and a few species common in the surrounding regrowth, notably *Dillenia nalagi*.

Wanigela—Cape Vogel

Two alliances of disclimax grassland, the *Saccharum spontaneum/Imperata cylindrica* alliance and the *Themeda australis/Alloteropsis semialata* alliance, are recognized by B. W. Taylor.[218] The first of the alliances has been subdivided into four related sub-alliances. These grassland areas were originally covered by forest. They are the result of intensive native gardening and are being maintained by annual hunting fires. Each alliance and sub-alliance consists of a series of associations, the particular association present being dependent on the effect of the burning, the taller occurring where the effect is less intense. Many factors can be correlated with the distribution of the associations through their effect on the severity of the burning, thus tall associations in the series tend to be found in areas of higher rainfall, in depressions or other low-lying areas, in grass stands of small size particularly if distant from native settlement, and on deep or more fertile soils, particularly those of finer texture. The short associations tend to be found on well to excessively drained sites, on eroded and shallow soils, and in areas with a marked dry season, particularly in areas subject to frequent burning due to the proximity of villages.

Saccharum spontaneum/Imperata cylindrica Alliance

Sub-alliances:

(i) *Saccharum spontaneum/Imperata cylindrica/Ophiuros exaltatus.*
Mature stands 2–4 m. high. Most important associated species are the grasses *Sorghum nitidum, Apluda mutica* and *Coelorhachis rottboellioides*, with many species of Compositae, Cyperaceae, Leguminosae and Euphorbiaceae, with small shrubs and scattered trees.

(ii) *Saccharum spontaneum/Imperata cylindrica/Coelorhachis rottboellioides.*
Shorter stages in the series of associations cover more extensive areas.

(iii) *Imperata cylindrica/Ischaemum barbatum.*
Community ranges in height from 60 cm. to 2 m. and is mainly densely packed. Dominant species are *Saccharum spontaneum, Imperata cylindrica, Ischaemum barbatum, Alloteropsis semialata* and in a few areas *Phragmites karka.* This sub-alliance occurs on sites apparently once covered by (i), now flooded, frequently dry out and are regularly burnt.

(iv) *Saccharum spontaneum/Imperata cylindrica/Cymbopogon procerus.*
Community 2–4 m. high but somewhat open. Other grasses present are *Hymenachne amplexicaulis* and *Ischaemum barbatum.*

Themeda australis/Alloteropsis semialata Alliance

Continuous cover 60 cm. to 2 m. high before annual burning. The sequence of dominants in the alliance with increasing severity of burning is *Saccharum spontaneum, Coelorhachis rottboellioides* or *Ophiuros exaltatus, Imperata cylindrica, Themeda australis* and *Alloteropsis semialata.* On any larger stand, two or more of these species may occur as local dominants. Common associated grasses are *Sorghum nitidum, Eragrostis elongata, Eragrostis* cf. *brownii, Capillipedium parviflorum, Heteropogon contortus* and *Ischaemum barbatum.*

Port Moresby–Kairuku

MIXED HERBACEOUS VEGETATION

Types in which non-graminioid herbs are important. A heterogeneous assemblage—some are pioneer or seral vegetations, others occur in environments very marginal for plant growth, according to P. C. Heyligers.[90] Grasses present are species of *Spinifex*, *Thuarea*, *Remirea* and minor *Sporobolus*, *Digitaria*, *Apluda*, *Setaria* and *Imperata*; *Leersia*, *Hymenachne* and *Echinochloa* and scattered *Phragmites*; *Imperata*, *Heteropogon*, *Themeda australis*, *Rhynchelytrum*, *Eriachne* and *Saccharum spontaneum*.

GRASSLANDS

Vegetation types dominated by grasses and subdivided into low grassland up to 60 cm., mid-height up to 1·5 m. and tall more than 1·5 m. Scattered trees and shrubs occur and some areas grade into savannahs. Low grasslands are confined to salt marshes, mid-height grasslands and one type of tall grassland are maintained by repeated burning which prevents woody regrowth. Other types of tall grassland occur in seasonally or permanently inundated situations.

The vegetation types are named after one or two predominant species such as:

(i) low grassland—*Sporobolus virginicus*/*Eriochloa procera*, with *Chloris barbata*, *Pluchea indica*, *Imperata cylindrica*, *Themeda novoguineensis*, *Th. australis* and *Heteropogon*.

(ii) mid-height grassland—*Ophiuros exaltatus*/*Imperata cylindrica* with *Saccharum spontaneum*, *Themeda novoguineensis*; *Ophiuros exaltatus*/*Themeda australis*, with *Saccharum spontaneum*, *Imperata cylindrica* and species of *Capillipedium*, *Heteropogon* and *Phragmites*; *Imperata cylindrica*/*Themeda australis*.

(iii) tall grassland and grass vegetation—*Saccharum spontaneum*/*Imperata cylindrica*; *Phragmites karka*/*Saccharum robustum*; pure stands of *Saccharum robustum*.

SAVANNAHS

Open tree storey and a ground cover of grasses, structurally and to a certain degree floristically similar to the woodlands of northern Australia (see discussion of term *savannah* on p. 100).

(i) *Themeda australis/Eucalyptus* (*alba, confertiflora, papuana*)
(ii) *Themeda australis/Timonius*
(iii) *Ophiuros exaltatus/Eucalyptus alba*
(iv) *Ophiuros exaltatus/Eucalyptus papuana*
(v) *Ophiuros exaltatus/Eucalyptus tereticornis*
(vi) *Ophiuros/Timonius*
(vii) *Themeda novoguineensis/Eucalyptus*, with species of *Panicum, Arundinella, Imperata, Heteropogon, Eriachne* or *Eulalia*
(viii) *Imperata cylindrica/Eucalyptus* (*papuana* and/or *alba*), with scattered species of *Cymbopogon, Heteropogon, Eulalia, Panicum, Capillipedium*, or *Themeda novoguineensis*.

Wabag-Tari

The predominant climax vegetation is a dense rain forest which has developed under conditions of high and largely non-seasonal rainfall.[190] A broad altitudinal zonation is:

below 1,200 m.: lowland rain forest with three tree layers
900–3,000 m.: lowland montane rain forest with two tree layers
3,000–3,750 m.: montane rain forest with one tree layer
above 3,750 m.: alpine grasslands.

Three major types of vegetation are recognized:

Forests
Regrowth vegetation
Grassland, swamp and bog communities.

INDUCED GRASSLANDS

The short grasslands of the inhabited highland valleys are man-made communities dominated for the most part by lowland short-grass species of wide distribution, the results of long-term biotic interference, clearing, burning and gardening, on the original vegetation.

(i) *Capillipedium parviflorum* grassland, *Arundinella setosa* frequent co-dominant, associate species *Eulalia trispicata, Apluda mutica, Ischaemum barbatum, Sorghum nitidum,*

Digitaria sp., *Ophiuros exaltatus*, *Paspalum* spp., *Pogonatherum paniceum*, *Sporobolus*, *Setaria* and *Themeda*.

(ii) *Themeda australis* grassland, co-dominant *Arundinella setosa*.

(iii) *Ischaemum polystachyum* grassland.

(iv) *Imperata cylindrica* grassland.

PHRAGMITES SWAMP

Phragmites karka in association with *Ischaemum polystachyum*, *Leersia hexandra* and *Isachne*.

LEERSIA SWAMP

The whole vegetation is frequently a floating mass firm enough to walk over.

GRASS-SEDGE BOG

Characteristic grasses are *Dimeria dipteros*, *Isachne arfakensis*, *Paspalum* sp. and at higher elevations *Agrostis reinwardtii* and *Arundinella furva*.

LOWER MONTANE GRASSLAND

In this category are a number of discrete, mostly uninhabited grasslands occurring mainly in the Western Highlands. Typically occupying valleys at 2,400–2,700 m., these grasslands are considered to be the result of frost-pocket conditions.

Diagnostic here are the grasses *Arundinella furva* and *Imperata exaltata*. Others are *Agrostis reinwardtii*, *Muehlenbergia arisanensis*, *Dichelachne novoguineensis*, *Anthoxanthum angustum*, *Deyeuxia brassii*, *Deschampsia klossii*, *Hierochloe longifolia*, and sedges. Drainage channels traversing the grasslands are typically filled or lined with *Machaerina rubiginosa* (syn. *Cladium glomeratum*) sedge. The cover is often sparse with bare ground between the small tussocks. Small herbs and dwarf woody shrubs such as *Hypericum*, *Potentilla*, *Rhododendron*, *Styphelia*, *Gaultheria*, *Haloragis*, *Lycopodium* and the fern *Gleichenia*, are a feature. Tree ferns also occur.

ALPINE GRASSLAND

This type embraces not only the tussock grasslands of the mountain summit areas which are above the tree line, but also some extensive valley grasslands below the forest limits. The Sugarloaf and McNicoll plateaux at about 2,700 m. are examples where, due to cold-air drainage, tree growth has been locally inhibited and downward extensions of alpine tussock grassland from higher levels occur. These aspects are less endowed with associated alpine species than the true summit areas of Mt. Giluwe and The Sugarloaf. The alpine grasslands are predominantly composed of tussock grasses in thick clumps 22·5–30 cm. across and 60–90 cm. high. Particularly common here are *Danthonia archboldii*, *Danthonia vestita*, *Dichelachne novoguineensis*, *Hierochloe longifolia*, *Deschampsia klossii*, *Anthoxanthum angustum*, *Deyeuxia brassii*, *Festuca papuana* and *Poa crassicaulis*.

Small shrubs such as *Styphelia*, *Rhododendron*, *Vaccinium*, *Haloragis*, *Detzneria*, *Gaultheria*, *Anaphalis*, *Keysseria* and *Coprosma* are present and in some places almost warrant an additional vegetation category of 'alpine heath'. Between the tussocks many low or prostrate herbs occur, including *Ranunculus*, *Gentiana*, *Potentilla*, *Lycopodium*, *Euphrasia*, *Drapetes*, *Oreomyrrhis*, the ferns *Papuapteris linearis* and *Gleichenia vulcanica*, and mosses. Numerous tree ferns, their gaunt forms rising from 60 cm. to 3·5 m., often give these grasslands an unusual local aspect, whence the popular name 'tree-fern grassland'. Small sedge bogs occur in wet sites.

Chapter XIII
AUSTRALIA

Phytogeography and migration

Continuing the theme of migration and relations between regions, we find that eastern Australia has been a very important migration route for a very long period (Nancy Burbidge),[24] being linked with the northern hemisphere through Malesia (Sumatra track, Luzon track, New Guinea track).[229, 230, 231] The northward migration of Australian elements has apparently been less successful than the southward migration of Malesian elements. The ecological barrier formed by dense tropical communities must have inhibited the northward movement of the light-requiring Australian types. The significance of such a barrier would be dependent on the climatic conditions during periods of land continuity.

According to R. L. Specht[208] the Indo-Malayan element has been isolated in Australia since early Tertiary times when it migrated to that continent across Torres Strait by means of island links to the north. From here it spread westward and southward to mingle with the Antarctic elements which invaded Australia via a land bridge from the south. From these two floras the characteristic Australian element arose. During the early Tertiary period the three elements became widespread across much of Australia, which then experienced a cool subtropical climate in the south and probably a tropical climate in the north.

Much of the migration into Australia was from the north under such a climatic regime.[208] Apart from this strong southward migration of the Indo-Malesian flora some tropical Australian genera, including *Eucalyptus*, *Grevillea*, *Casuarina* and others, gave rise to the occasional species which were able to find ecological niches in the predominantly tropical environment to the north of Australia. All these minor representatives of the typical Australian flora indicate a weak outward migration from Australia through New Guinea along the various inter-island linkages of the Malesian Archipelago into Asia during the Upper Cretaceous-Eocene periods.

The land bridges had disappeared by the Oligocene leaving behind a greatly intermixed flora which became dispersed during the warmer Miocene-Pliocene periods. The Pleistocene ushered in a period of great climatic instability which has continued until today. Successive periods of aridity and humidity markedly affected the vegetation of Australia as the mean path of the low-pressure systems was displaced latitudinally with the repetitive formation and later melting of the polar ice-caps. Drier conditions in the north were associated with a fall in the sea level sufficient to create a land bridge between Australia and New Guinea. Some species could invade the coastal fringe of New Guinea, and a few New Guinean species could penetrate across Torres Strait and into Arnhem Land. Even these species had been isolated in New Guinea from the rest of the Indo-Malesian vegetation since the Eocene period.

Climatic stress during periods of aridity was so severe that certain species and communities lost their original continuous distribution and became broken into an infinite number of disjunct communities. A number of Indo-Malesian species which must have become widespread in Australia since their original invasion in the Upper Cretaceous-Eocene periods are now recorded in only isolated stands; others retain their widespread distribution across Northern Australia. Many of the original invaders are still preserved, with speciation insufficient to cause the extinction of the original tropical flora. Specht concludes: 'In all, it must be remembered that the climatic stresses of the north occurred at different times from those in the south of the continent. For instance, the extreme aridity of southern Australia during the Post-Glacial Optimum was not a period of aridity in the north. Northern Australia is suffering under drier conditions now than it experienced then. The vegetation patterns seen today in the north are largely an expression of the Pleistocene changes in climate, and not of those of the Recent.'[208]

Old species have had time and opportunity to become established over large areas (J. G. Wood);[259] S. T. Blake has pointed out that several species are so widespread over northern Australia as to suggest a definite vegetation region, a view supported by the similar pattern of distribution of several genera of grasses (Andropogoneae) and such genera as *Brachychiton*, *Ficus*, *Fimbristylis*, and such a family as the Combretaceae. A fact to be stressed is that the Australian plant communities are young, yet relatively stable, and that their distribution within a climatic zone has been determined chiefly by edaphic factors; other factors have of course also played a part, e.g. individual dispersal capacity, chance dispersals, location of surviving centres, degrees of biotypical differentiation, and ecological barriers. The basis underlying ecology is a physiological one, concerned with individual tolerances of species. In the case of species growing together in a community, the potential environment of the individual species overlaps the potential environment. The present-day distribution and grouping into communities are the result of past history associated with often extreme sensitivity to changes in the micro-habitat, the boundaries of which are not usually abrupt but present a varying continuum. The character of the community depends primarily upon the growth tolerances of the component species. The most important work confronting ecologists is the determination of these tolerances.

Grasses of Arnhem Land

In a chapter on climate, geology, soils and plant ecology in the Records of the American-Australian Expedition, R. L. Specht[207] discusses the vegetation under the following units (following R. L. Crocker and J. G. Wood[58] rather than C. S. Christian and G. A. Stewart).[46]

Monsoon forest formation
Tall open forest
Quartzite and sandstone edaphic complex
Sandy fan delta edaphic complex
Savannah woodland formation (*Eucalyptus tectifica* association) on granitic hills
Freshwater stream, swamp and marsh edaphic complex
Estuarine (black soil) plain edaphic complex
Mangrove forest edaphic complex
Coastal dune edaphic complex.

Only one grass species, *Eriachne mucronata*, is mentioned in the savannah woodland with the small twisted box eucalypt. By far the greater number of grass species are reported in the freshwater stream, swamp and marsh edaphic complex, the area of the largely ephemeral streams that arise in the torrential monsoon rains. These low-lying areas are all subject to waterlogging for 6–12 months of the year, and the duration largely governs the composition of the plant communities that occur.

In Table 28 is given the distribution throughout Australia and overseas[208] of every species which has been recorded in the Arnhem Land aboriginal reserve. Doubtful localities have been left blank. A cross indicates the presence, a dash the absence of the species in that particular environment. The numbers at the top of each column refer to the following localities:

Australia
- (1) Kimberley and Hamersley Range regions, Western Australia
- (2) Victoria River Downs region, Northern Territory
- (3) Darwin-Katherine region, Northern Territory
- (4) Arnhem Land, Northern Territory
- (5) Southern Gulf of Carpentaria, Northern Territory and Queensland
- (6) Cape York region, Queensland
- (7) Southern Coastal region, Queensland, and Northern Coastal region, New South Wales
- (8) Coastal New South Wales
- (9) Southern Australia
- (10) Interior

Overseas
- (1) New Guinea
- (2) Malesian Archipelago
- (3) Philippine Islands
- (4) Continental S.E. Asia
- (5) Indian Peninsula
- (6) East African Islands
- (7) Tropical Africa
- (8) Melanesia and Micronesia
- (9) Polynesia
- (10) Tropical America

TABLE 28

Distribution throughout Australia and overseas of grass species recorded for Arnhem Land aboriginal reserve[208]

	Australia										Overseas									
	1	2	3	4	5	6	7	8	9	10	1	2	3	4	5	6	7	8	9	10
PANICOIDEAE																				
Alloteropsis semialata (R.Br.) Hitchc.	×	×	×	×	×	×	×	–	–	–	×		×	×	×		×	–	–	–
Brachiaria argentea (R.Br.) Hughes	–	–	–	×	–	–	–	–	–	–	–	–	–	–	–	–	–	–	–	–
B. holosericea (R.Br.) Hughes	×	×	×	×	×	×	–	–	–	×	×	–	–	–	–	–	–	–	–	–
B. mutica (Forsk.) Stapf	–			×											×		×			
B. polyphylla (R.Br.) Hughes	–	–	×	×	–	–	–	–	–	–	–	–	–	–	–	–	–	–	–	–
Cenchrus brownei Roem. & Schult.	–	–	–	×	–	–	–	–	–	–	×	×	×	×	–	–	–	–	–	×
Chamaeraphis hordeacea R.Br.	–	–	–	×	–	–	–	–	–	–	–	–	–	–	–	–	–	–	–	–
Digitaria adscendens (H.B.K.) Henr.	×	–	–	×	–	×	×	×	×	–	×	×	×	×	×	×	×	×	×	×
D. ctenantha (F. Muell.) Hughes	×(H)	×	–	×	–	–	–	–	–	×	–	–	–	–	–	–	–	–	–	–
D. gibbosa (R.Br.) Beauv.	–	×	×	×	–	×	–	–	–	–	–	–	–	–	–	–	–	–	–	–
D. leucostachya (Domin) Henr.	–	–	–	×	–	–	×	–	–	–	–	–	–	–	–	–	–	–	–	–
D. longiflora (Retz.) Pers.	×	×	–	×	–	–	×	×	–	–	×	×	×	×	×	×	×	–	–	–
D. papposa (R.Br.) Beauv.	–	–	–	×	×	–	–	–	–	–	–	–	–	–	–	–	–	–	–	–
Echinochloa colonum (L.) Link	×	×	×	×	–	–	–	–	–	×	×	×	×	×	×	×	×	×	×	×
Eriochloa crebra S. T. Blake	–	–	×	×	–	–	–	–	–	–	–	–	–	–	–	–	–	–	–	–
Ichnanthus pauciflorus (R.Br.) Hughes	–	–	–	×	×	–	–	–	–	–	–	–	–	–	–	–	–	–	–	–
Isachne confusa Ohwi	–	–	–	×	–	×	–	–	–	–	×	×	–	–	–	–	–	–	–	–
Panicum airoides R.Br.	–	–	×	×	×	×	–	–	–	×	–	–	–	–	–	–	–	–	–	–
P. cymbiforme Hughes	×	–	–	×	–	–	–	–	–	–	–	–	–	–	–	–	–	–	–	–
P. decompositum R.Br.	×	×	×	×	–	×	×	×	–	×	–	–	–	–	–	–	–	–	–	–
P. delicatum Hughes	×	–	×	×	–	–	–	–	–	–	–	–	–	–	–	–	–	–	–	–
P. seminudum Domin	–	–	–	×	–	×	–	–	–	–	–	–	–	–	–	–	–	–	–	–
P. trachyrhachis Benth.	–	×	×	×	–	×	×	–	–	–	–	–	–	–	–	–	–	–	–	–
P. trichoides Sw.	–	–	×	×	–	×	–	–	–	–	–	×	×	–	–	–	–	–	–	×
Paspalidium distans (Trin.) Hughes	×	–	×	×	–	×	×		–	–	×									
P. gracile (R.Br.) Hughes	×	–	–	×	–	×	×	×	×	×	–	–	–	–	–	–	–	–	–	–
P. rarum (R.Br.) Hughes	×	–	–	×	–	–	–	–	–	×	–	–	–	–	–	–	–	–	–	–
Paspalum orbiculare G. Forst.	–	×	×	×	×	×	×	×	–	–	×	×	×	×	×	×	×	–	–	–
Pseudoraphis spinescens (R.Br.) J. Vick.	×	–	×	×	×	×	×	×	×	–	–	×	×			–	–	–	–	–
Sacciolepis indica (L.) Chase	×	×	×	×	×	×	×	×			×									

	Australia 1	2	3	4	5	6	7	8	9	10	Overseas 1	2	3	4	5	6	7	8	9	10
Schult.	–	–	–	×	–	×	–	–	–	–	×	×	×	×	×	×	–	–	×	–
Xerochloa barbata R.Br.	×	–	–	×	×	–	–	–	–	–	–	–	–	–	–	–	–	–	–	–
X. imberbis R.Br.	×	×	×	×	–	–	–	–	–	–	–		–	–	–		–	–	–	–
ANDROPOGONOIDEAE																				
Arundinella nepalensis Trin.	×	×	×	×	–	×	×	×	–	–	×	×	×	×	×	×	–	–	–	–
Capillipedium parviflorum (R.Br.) Stapf	–	–	×	×	–	×	×	×	–	×	–	–	–	–	–	–	–	–	–	–
Chionachne cyathopoda (F. Muell.) F. Muell. ex. Benth.	×	×	×	×	×	×	×	–	–	–	–	–	–	–	–	–	–	–	–	–
Chrysopogon pallidus (R.Br.) Trin. ex Steud.	×	–	–	×	–	–	–	–	–	–	–	–	–	–	–	–	–	–		–
C. setifolius Stapf	–	–	×	×	–	–	–	–	–	–	–	–	–	–	–	–	–	–	–	–
Coelorhachis rottboellioides (R.Br.) A. Camus	×	×	×	×	×	×	–	–	–	–	×	–	–	–	–	–	–	–	–	–
Cymbopogon exaltatus (R.Br.) Domin	×(H)	×	×	×	–	×	–	–	×	×	–	–	–	–	–	–	–	–	–	–
C. procerus (R.Br.) Domin	×	×	×	×	–	–	–	–	–	×	×	–	–	–	–	–	–	–	–	–
C. refractus (R.Br.) A. Camus	–	–	–	×	–	×	×	×	×	–		×	–	–	–	–	–	–	×	–
Dichanthium superciliatum (Hack.) A. Camus	×		×	×							–	×	–	–	–	–	–	–		–
Elyonurus citreus Munro ex Benth.	–	–	–	×	–	×	–	–	–	–	×	–	–	–	–	–	–	–	–	–
Eulalia fulva (R.Br.) Kuntze	×	–	×	×	–	×	×	–	–	×		×	×	×	×	–	–	–	–	–
E. mackinlayi (F. Muell. ex Benth.) S. T. Blake	–	–	×	×	–	–	–	–	–	–	–	–	–	–	–	–	–	–	–	–
aff. *Eulalia*				×																
Heteropogon contortus (L.) Beauv. ex Roem. & Schult.	×	×	×	×	×	×	×	–	–	–	×	×	×	×	×	×	×	×	×	×
H. triticeus (R.Br.) Stapf	–	–	×	×	–	×	–	–	–	–	×	×	×	×	×	–	–	–	–	–
Imperata cylindrica (L.) Beauv. var. *major* (Nees) C. E. Hubb.	×	×	×	×	×	×	×	×	×	×	×	×	×	×	×	×	×	×	×	×
Ischaemum arundinaceum F. Muell. ex Benth.	×	–	×	×	–	–	–	–	–	–	×	–	–	–	–	–	–	–	–	–
I. decumbens Benth.	–	–	×	×	–	–	–	–	–	–	–	–	–	–	–	–	–	–	–	–
I. sp.				×																
Iseilema vaginiflorum Domin	×			×						×										
Ophiuros exaltatus (L.) Kuntze	×	×	–	×	×	×	–	–	–	–	×	×	×	×	×	×	×	–	–	–
Pseudopogonatherum contortum (Brogn.) A. Camus	–	–	×	×	–	×	–	–	–	–		×	×	×	×	–	–	–	–	

TABLE 28—continued

	Australia 1	2	3	4	5	6	7	8	9	10	Overseas 1	2	3	4	5	6	7	8	9	10
P. irritans (R.Br.) A. Camus	–	–	–	×	–	×	–	–	–	–	×	–	–	–	–	–	–	–	–	–
Rottboellia exaltata L.f.	–	–	–	×	–	–	–	–	–	–	×	×	×	×	×	–	×	–	–	–
R. formosa R.Br.	×	×	×	×	–	×	–	–	–	–	–	–	–	–	–	–	–	–	–	–
Schizachyrium spp.			×	×																
Sclerandrium grandiflorum S. T. Blake	–	–	×	×	–	×	–	–	–	–	–	–	–	–	–	–	–	–	–	–
S. truncatiglume (F. Muell. ex Benth.) Stapf & C. E. Hubb.	×	–	×	×	–	–	–	–	–	–	×	–	–	–	–	–	–	–	–	–
Sehima nervosum (Rottb.) Stapf	×	×	×	×	×	×	–	–	–	×	–	×	–	×	×	×	×	–	–	–
Sorghum intrans F. Muell. ex Benth.	×	–	×	×	×	–	–	–	–	–	–	–	–	–	–	–	–	–	–	–
S. plumosum (R.Br.) Beauv.	×	×	×	×	–	×	×	×	×	×	–	×	–	–	–	–	–	–	–	–
S. spp.				×																
Thaumastochloa sp.	–	–	×	×	–	×	–	–	–	–	–	–	–	–	–	–	–	–	–	–
Themeda arguens (L.) Hack.	×	–	×	×	–	×	–	–	–	–	×		×	×	–	–	–	–	–	–
Vetiveria elongata (R.Br.) Stapf ex C. E. Hubb.	–	–	×	×	×	×	–	–	–	–	×	–	–	–	–	–	–	–	–	–
V. pauciflora S. T. Blake	×	–	×	×	×	×	–	–	–	–	–	–	–	–	–	–	–	–	–	–
ORYZOIDEAE																				
Oryza fatua Koen. ex Trin.	×	×	×	×	×	–	–	–	–	–										
ERAGROSTOIDEAE																				
Aristida browniana Henr.	×	×	×	×	×	–	–	–	–	×	–	–	–	–	–	–	–	–	–	–
A. hygrometrica R.Br.	×	×	–	×	×	×	–	–	–	–	–	–	–	–	–	–	–	–	–	–
Brachyachne convergens (F. Muell.) Stapf	×	×	–	×	–	–	–	×	–	–	–	–	–	–	–	–	–	–	–	–
B. tenella (R.Br.) C. E. Hubb.	×	×	–	×	×	–	–	–	–	–	–	–	–	–	–	–	–	–	–	–
Chloris pumilio R.Br.	×	–	–	×	×	–	–	–	–	–	–	–	–	–	–	–	–	–	–	–
Cynodon arcuatus Presl	–	–	×	×			–	–	–	–	–	–	×	–	–	–	–	–	–	–
C. dactylon (L.) Pers.	×	–	×	×	–	×	×	×	×	×	×	×	×	×	×	×	×	×	×	×
Dactyloctenium radulans (R.Br.) Beauv.	×	×	–	×	×	×	–	–	–	×	–	–	–	–	–	–	–	–	–	–
Diplachne parviflora (R.Br.) Benth.	×	–	–	×	–	–	–	–	–	–	–	–	–	–	–	–	–	–	–	–
Ectrosia agrostoides Benth.	–	–	–	×	–	–	–	–	–	–	–	–	–	–	–	–	–	–	–	–
E. leporina R.Br. var. *leporina* C. E. Hubb.	–	×	×	×	×	×	×	–	–	×	–	–	–	–	–	–	–	–	–	–
E. leporina R.Br. var. *micrantha* Benth.	–	×	–	×	×	–	–	–	–	–	–	–	–	–	–	–	–	–	–	–
E. schultzii Benth.	×	×	×	×	–	–	–	–	–	–	–	–	–	–	–	–	–	–	–	–
Eleusine indica (L.) Gaertn.	×	–	×	×	–	×	×	×	–	–	×	×	×	×	×	×	×	×	×	×
Eragrostis bleeseri Pilger	–	–	×	×	–	–	–	–	–	–	–	–	–	–	–	–	–	–	–	–

	Australia 1	2	3	4	5	6	7	8	9	10	Overseas 1	2	3	4	5	6	7	8	9	10
E. spp.				×																
Heterachne abortivum (R.Br.) Druce	–	–	×	×	×	–	–	–	–	–	–	–	–	–	–	–	–	–	–	–
Leptochloa brownii C. E. Hubb.	×	×	–	×	×	×	–	–	–	–		×			×					
Lepturus repens (Forst.) R.Br.	–	–	×	×	–	×	–	–	–	–	×	×	×	×	×	–	–	×	×	–
Microchloa setacea R.Br.	–	–	–	×	–	–	–	–	–	–					×		×			×
Perotis rara R.Br.	×	–	×	×	×	×	–	–	–	×	×		×	–	–	–	–	–	–	–
Sporobolus australasicus Domin	×			×																
S. pulchellus R.Br.	×	×	×	×	–	×	–	–	–	×		×			×	–	–	–	–	–
S. virginicus (L.) Kunth *sens. lat.*	×	×	×	×	–	×	×	×	–	×		×	×	×	×	×	×	×	×	×
Triraphis mollis R.Br.	×	–	×	×	–	×	×	×	–	×	–	–	–	–	–	–	–	–	–	–
Festucoideae																				
Astrebla squarrosa C. E. Hubb.	×	–	–	×	–	–	–	–	–	×	–	–	–	–	–	–	–	–	–	–
Enneapogon pallidus (R.Br.) Beauv.	×	–	–	×	×	×	–	–	–	×	–	–	–	–	–	–	–	–	–	–
Eriachne agrostidea F. Muell.	–	–	×	×	–	–	–	–	–	–	–	–	–	–	–	–	–	–	–	–
E. avenacea R.Br.	–	×	×	×	–	×	–	–	–	–	–	–	–	–	–	–	–	–	–	–
E. basedowii Hartley	–	–	–	×	–	–	–	–	–	–	–	–	–	–	–	–	–	–	–	–
E. burkittii Jansen	–	–	×	×	–	–	–	–	–	–	–	–	–	–	–	–	–	–	–	–
E. capillaris R.Br.	–	–	–	×	–	–	–	–	–	–	–	–	–	–	–	–	–	–	–	–
E. ciliata R.Br.	×	–	×	×	×	×	–	–	–	–	–	–	–	–	–	–	–	–	–	–
E. filiformis Hartley	–	–	–	×	–	×	–	–	–	–	–	–	–	–	–	–	–	–	–	–
E. glauca R.Br.	×	×	–	×	–	–	–	–	–	–	–	–	–	–	–	–	–	–	–	–
E. melicacea F. Muell.	×	×	–	×	–	–	–	–	–	×	–	–	–	–	–	–	–	–	–	–
E. mucronata R.Br.	×	–	–	×	–	×	–	–	–	×	–	–	–	–	–	–	–	–	–	–
E. obtusa R.Br.	×	×	×	×	×	×	×	–	–	×	–	–	–	–	–	–	–	–	–	–
E. schultziana F. Muell.	–	–	×	×	–	–	–	–	–	–	–	–	–	–	–	–	–	–	–	–
E. setacea Benth.	–	–	–	×	–	–	–	–	–	–	–	–	–	–	–	–	–	–	–	–
E. squarrosa R.Br.	×	×	–	×	×	×	–	–	–	–	×	×	–	–	–	–	–	–	–	–
E. stipacea F. Muell. var. *hirsuta* Hartley	×	×	×	×	–	–	–	–	–	–	–	–	–	–	–	–	–	–	–	–
E. triodioides Domin	–	–	–	×	–	×	–	–	–	–	–	–	–	–	–	–	–	–	–	–
E. triseta Nees ex Steud.	–		×	×							×	×	×		×					
Phragmites karka (Retz.) Trin. ex Steud.	×	–	×	×	–	×	×	×	×	×	×	×	×	×	×	–	–	–	–	–
Plectrachne pungens (R.Br.) C. E. Hubb.	×	×	×	×	–	–	–	–	–	×	–	–	–	–	–	–	–	–	–	–
Triodia microstachya R.Br.	×	×	×	×	–	–	–	–	–	–	–	–	–	–	–	–	–	–	–	–
T. procera R.Br.	×	×	–	×	–	–	–	–	–	–	–	–	–	–	–	–	–	–	–	–
T. pungens R.Br.	×	–	–	×	×	–	–	–	–	×	–	–	–	–	–	–	–	–	–	–

TABLE 29

Frequencies of Arnhem Land species in other parts of Australia and the world[208]

	Per cent
Australia	
(1) Kimberley and Hamersley Range regions, Western Australia	45
(2) Victoria River Downs region, Northern Territory	39
(3) Darwin–Katherine region, Northern Territory	41
(4) Arnhem Land, Northern Territory	100
(5) Southern Gulf of Carpentaria, Northern Territory and Queensland	28
(6) Cape York region, Queensland	54
(7) Southern coastal region, Queensland, and northern coastal region, New South Wales	22
(8) Coastal New South Wales	10
(9) Southern Australia	5
(10) Interior	19
Overseas	
(1) New Guinea	32
(2) Malesian Archipelago	26
(3) Philippine Islands	22
(4) Continental S.E. Asia	24
(5) Indian Peninsula	26
(6) East African Islands	10
(7) Tropical Africa	14
(8) Melanesia and Micronesia	9
(9) Polynesia	11
(10) Tropical America	8

Plant communities of Western Queensland

In his paper on these communities, S. T. Blake[12] is referring to an area of about 330,000 square miles further south, embracing the greater part of purely pastoral Queensland, excluding the very large area around the Gulf of Carpentaria which constitutes a separate botanical province. Forty-five plant communities are arranged in fourteen groups and described as to habitat, dominant species, floristic detail, reactions to stocking, interrelations and history. The whole of the area lies within the 750 mm. isohyet and the greater part within the 500 mm. isohyet (down to 150 mm. in the far south-west). Rainfall is erratic throughout, there are occasional years of heavy rainfall and frequent periods of prolonged drought; rainfall is concentrated in summer in the north, while winter rains increase in importance towards the south-east.

Blake recognizes three well-marked groups of grasslands, named according to the dominant genus.

(*a*) Blue grass country with *Dichanthium sericeum* (blue grass) dominant,

(*b*) Mitchell grass country with *Astrebla* spp. dominant, and

(*c*) *Triodia* grasslands.

In the better rainfall areas and generally on black earth, the grass species are tufted, the tufts being very leafy, fairly close together and not very large. The more prominent species are shorter-lived and more fibrous than the dominants of the other types of grassland. They are numerous and frequently two or more are co-dominant. The most characteristic are blue grass (*Dichanthium sericeum* and certain closely allied forms), *Bothriochloa erianthoides, Paspalidium globoideum, Panicum decompositum, P. queenslandicum, Digitaria divaricatissima* and *Thellungia advena* (water grass). Other characteristic plants are *Caspedia uniflora, Ixiolaena brevicompta, Sida pleiantha* and species of *Indigofera* and *Neptunia.*

There are two climatic subtypes:

The southern zone lies chiefly in the Darling Downs district, where winter rainfall is pronounced. The genera *Danthonia* and *Stipa* and the species *Aristida leptopoda* are important, with *Themeda avenacea* in places.

In the northern zone, *Stipa* and *Danthonia* are absent, while *Astrebla* and *Iseilema* occur, usually sparsely and often as the result of invasion. *Aristida leptopoda* is common in certain places, perhaps introduced by stock from the south.

On small bare spaces are to be found such species as *Brachyachne convergens* and *B. tenella, Enneapogon nigricans, Tragus biflorus* and a few other grasses, together with *Portulaca intraterranea, Trianthema cristallina, Tribulus terrestris, Atriplex semibaccata, Euphorbia drummondii* and *Boerhavia diffusa.* With degradation the comparatively worthless *Panicum decompositum* comes in, often with *Aristida leptopoda.*

S. T. Blake (1964) has provided information on the coastal and sub-coastal regions in Queensland, supplementary to his paper[12] discussed above. Apart from areas of artificial pastures, these include the cleared, partly cleared and uncleared *Eucalyptus* forest areas discussed below, and

(*a*) large areas along and close to the coast, some of them flooded at spring tides, others now permanently above tide mark, dominated by *Sporobolus virginicus* (marine couch, salt-water couch);

(*b*) local areas of clay soil north of the Tropic of Capricorn dominated by *Ophiuros exaltatus*; the total area appears to be small;

(*c*) in Cape York Peninsula (in the broad sense) and parts of the Northern Territory there are communities dominated by *Dichanthium fecundum*;

(*d*) local swampy areas, some of which dry out in dry times; *Paspalum distichum, Ischaemum, Oryza, Leersia, Panicum paludosum* and other grasses are found, together with Cyperaceae, etc.; there is considerable variation in size and composition of these communities.

The Division of Land Research has conducted one of its surveys in western Queensland, in the Leichhardt-Gilbert area.[53] With a wide range of habitats due to wide variations in rainfall and topography, it is found that the vegetation is equally diverse, comprising several hundred associations in the sense of N. C. W. Beadle and A. B. Costin.[7] The common feature that probably has a selective effect on the physiognomy of the vegetation is the long annual drought, most of the rain falling in four or five summer months. Hence the vegetation is mostly grasslands, or woodlands with grassy under-storeys. Grasslands occupy about one-third of the area. According to R. A. Perry and M. Lazarides,[53] the dominant grasses are mid-height perennials (0·5–1 m.)—*Astrebla* on the cracking clay plains, *Aristida* most widespread; the prominent grasses in the moderate to high rainfall areas are *Dichanthium fecundum, D. sericeum, Bothriochloa intermedia, B. ewartiana, B. decipiens* var. *cloncurrensis, Themeda australis, Heteropogon contortus* and *Chrysopogon fallax*.

One of the first Land Research Reports deals with the Barkly region, which fits partly into Queensland and partly into Northern Territory along the Gulf of Carpentaria.[51] The heavy-soil plains that occur in areas of higher rainfall have pastures inferior to the typical *Astrebla* pastures. The zone of transition runs along the 450 mm. isohyet on the tableland and the 600 mm.

isohyet in Gregory River valley. With increasing rainfall there is a tendency for species of *Eulalia*, *Bothriochloa* and *Dichanthium* to replace the *Astrebla* spp. as dominants. These are the *Eulalia fulva/Dichanthium fecundum* grasslands, associated with woodlands dominated by *Eucalyptus microtheca* and *Bauhinia cunninghamii*.

Northern Territory

R. A. Perry has brought together all available experience on the pasture lands of the Northern Territory in Land Research Series Bulletin No. 5.[169] This is a complete review of the environment (physiography, climate, soils, vegetation, fires, animals), the pastoral industry (management, markets and transport, potential) and of the eight types of pasture lands that he recognizes in the territory. The following pasture types are relevant in the present context, out of the many that are given (see also Land Research Series Report on Tipperary).[54]

(i) Barley Mitchell grass. *Astrebla pectinata* with *Eulalia fulva* and species of *Dichanthium* and *Bothriochloa*.

(ii) Barley Mitchell grass and other perennial grasses. *Astrebla pectinata*, *Panicum whitei*, *P. decompositum*, *Dichanthium fecundum*, *Bothriochloa intermedia*, *Aristida latifolia*, *Astrebla squarrosa*, *Eulalia fulva*, *Chrysopogon fallax*, *Heteropogon contortus*, *Sehima nervosum*, and between the tussocks *Iseilema* spp.

(iii) Inferior Mitchell grass and other perennial grasses. *Astrebla squarrosa*, *Dichanthium fecundum*, *D. superciliatum*, *Bothriochloa ewartiana*, *B. intermedia*, *Ophiuros exaltatus*, *Panicum* spp., *Chrysopogon fallax*, *Aristida latifolia*, *Eulalia fulva* and *Arundinella nepalensis*.

(iv) Soft Spinifex plains. *Triodia pungens*, *Plectrachne schinzii*, *Aristida pruinosa*, *Cymbopogon bombycinus*.

(v) Semi-arid upland. *Aristida pruinosa*, *Themeda australis*, *Sehima nervosum*, *Chrysopogon pallidus*, *Cymbopogon bombycinus*.

(vi) Higher rainfall upland. *Sorghum plumosum*, *Themeda australis*, *Chrysopogon fallax*, *C. latifolius*, *Sehima nervosum*, *Heteropogon contortus*, *Aristida pruinosa*, *Brachyachne convergens*.

(vii) Lowlands interspersed with inaccessible hills. *Themeda australis*, species of *Dichanthium*, *Bothriochloa* and *Chrysopogon*, *Sehima nervosum*, *Brachyachne convergens*.

(viii) Annual sorghum and other tall grasses. *Sorghum intrans*, *S. stipoideum*, *Heteropogon triticeus*, *Sorghum plumosum*, *Chrysopogon latifolius* and *Coelorhachis rottboellioides*.

Western Australia

Land Research Series Reports Nos. 4 and 9 relate to the lands and pastoral resources of the North Kimberley area and to the lands of the West Kimberley area respectively.[129, 210] In the former, M. Lazarides recognizes fifteen pasture types classified into four groups: good, moderate, poor and very poor. The only type coming into the first category are the blue grass pastures, with the dominant species *Dichanthium fecundum*, *D. superciliatum*, *D. sericeum*, *D. annulatum*, *Bothriochloa ewartiana* and *B. intermedia*, with the associated annuals in the depressions of the gilgais: *Brachyachne convergens*, *Panicum decompositum*, *Eragrostis japonica*, species of *Paspalidium* and *Brachiaria*, *Eriachne glauca*, *Elytrophorus spicatus* and *Iseilema vaginiflorum*.

The native pastures of the West Kimberley area, according to N. H. Speck, have developed under an average rainfall of 400–1,000 mm. and in many respects have affinities with those of both arid and higher rainfall parts of northern Australia. N. H. Speck and M. Lazarides[210] have compiled Table 30 to show the various combinations of upper-storey trees and ground-storey grasses that characterize the vegetation of West Kimberley. Twenty-one different ground storeys have been recognized—their main characteristics have been summarized in Table 31. A few occur only as grasslands, some as grasslands and as grassy ground storeys to woodlands, and others only as grassy ground storeys to one or more tree communities.

Further information on the grass cover types of Northern and Western Australia can be found in the reports of the integrated surveys conducted by the Division of Land Research, CSIRO, e.g. Katherine/Darwin, Wiluna/Meekatbarra, West Kimberley.[46, 169, 210]

TABLE 30

Combinations of upper storeys and ground storeys in West Kimberley[210]

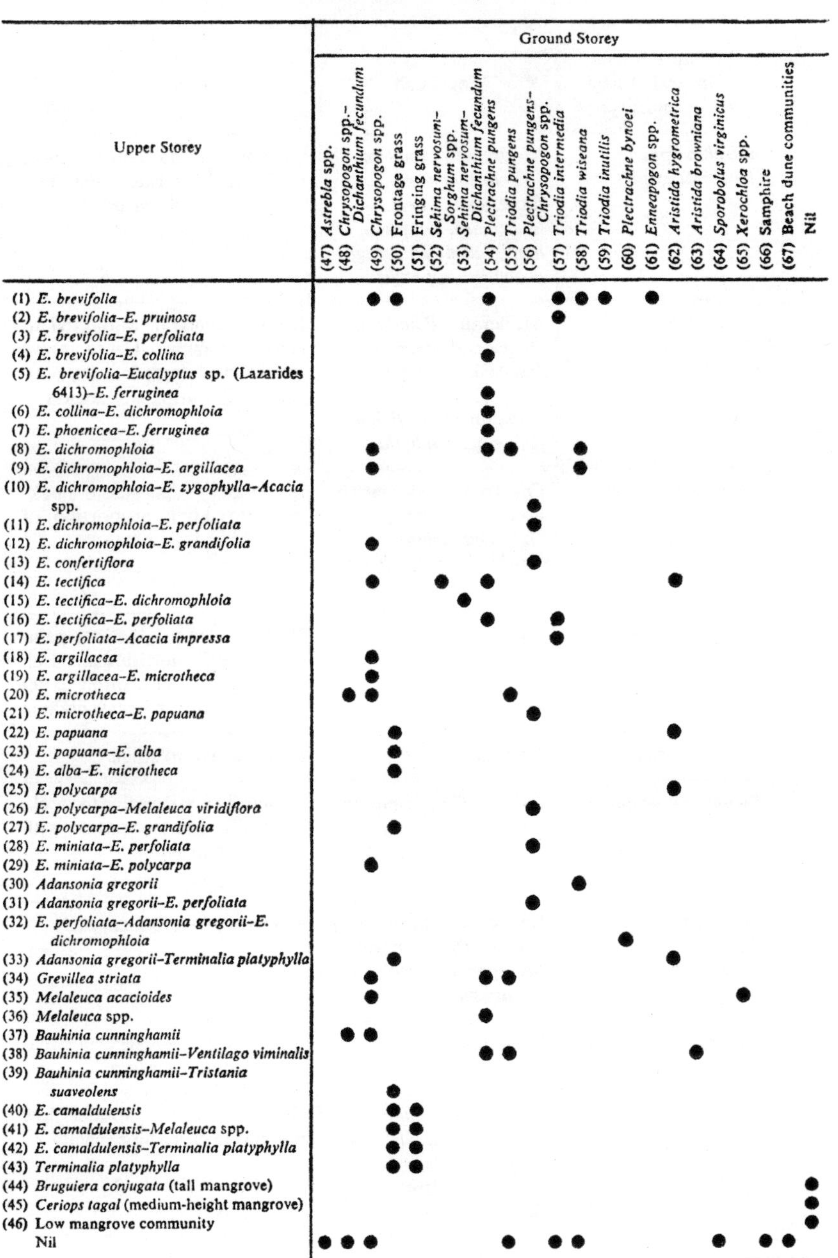

Upper Storey	Ground Storey																					
	(47) *Astrebla* spp.	(48) *Chrysopogon* spp.–*Dichanthium fecundum*	(49) *Chrysopogon* spp.	(50) Frontage grass	(51) Fringing grass	(52) *Sehima nervosum*–*Sorghum* spp.	(53) *Sehima nervosum*–*Dichanthium fecundum*	(54) *Plectrachne pungens*	(55) *Triodia pungens*	(56) *Plectrachne pungens*–*Chrysopogon* spp.	(57) *Triodia intermedia*	(58) *Triodia wiseana*	(59) *Triodia inutilis*	(60) *Plectrachne bynoei*	(61) *Enneapogon* spp.	(62) *Aristida hygrometrica*	(63) *Aristida browniana*	(64) *Sporobolus virginicus*	(65) *Xerochloa* spp.	(66) Samphire	(67) Beach dune communities	Nil
(1) *E. brevifolia*			●	●				●			●	●	●		●							
(2) *E. brevifolia–E. pruinosa*											●											
(3) *E. brevifolia–E. perfoliata*								●														
(4) *E. brevifolia–E. collina*								●														
(5) *E. brevifolia–Eucalyptus* sp. (Lazarides 6413)–*E. ferruginea*								●														
(6) *E. collina–E. dichromophloia*								●														
(7) *E. phoenicea–E. ferruginea*								●														
(8) *E. dichromophloia*			●					●	●			●										
(9) *E. dichromophloia–E. argillacea*			●									●										
(10) *E. dichromophloia–E. zygophylla–Acacia* spp.										●												
(11) *E. dichromophloia–E. perfoliata*										●												
(12) *E. dichromophloia–E. grandifolia*			●																			
(13) *E. confertiflora*										●												
(14) *E. tectifica*			●			●		●								●						
(15) *E. tectifica–E. dichromophloia*							●															
(16) *E. tectifica–E. perfoliata*								●			●											
(17) *E. perfoliata–Acacia impressa*											●											
(18) *E. argillacea*			●																			
(19) *E. argillacea–E. microtheca*			●																			
(20) *E. microtheca*		●	●						●													
(21) *E. microtheca–E. papuana*										●												
(22) *E. papuana*				●												●						
(23) *E. papuana–E. alba*				●																		
(24) *E. alba–E. microtheca*				●																		
(25) *E. polycarpa*																●						
(26) *E. polycarpa–Melaleuca viridiflora*										●												
(27) *E. polycarpa–E. grandifolia*				●																		
(28) *E. miniata–E. perfoliata*										●												
(29) *E. miniata–E. polycarpa*			●																			
(30) *Adansonia gregorii*												●										
(31) *Adansonia gregorii–E. perfoliata*										●												
(32) *E. perfoliata–Adansonia gregorii–E. dichromophloia*														●								
(33) *Adansonia gregorii–Terminalia platyphylla*				●												●						
(34) *Grevillea striata*			●					●	●													
(35) *Melaleuca acacioides*			●																●			
(36) *Melaleuca* spp.								●														
(37) *Bauhinia cunninghamii*		●	●																			
(38) *Bauhinia cunninghamii–Ventilago viminalis*								●	●								●					
(39) *Bauhinia cunninghamii–Tristania suaveolens*				●																		
(40) *E. camaldulensis*				●	●																	
(41) *E. camaldulensis–Melaleuca* spp.				●	●																	
(42) *E. camaldulensis–Terminalia platyphylla*				●	●																	
(43) *Terminalia platyphylla*				●	●																	
(44) *Bruguiera conjugata* (tall mangrove)																						●
(45) *Ceriops tagal* (medium-height mangrove)																						●
(46) Low mangrove community																						●
Nil	●	●	●						●		●	●						●		●	●	

Table 31

Main characteristics of the ground storeys (pasture types) in West Kimberley[210]

Pasture Type (Ground-storey Community)	Top Feed	Remarks
Astrebla spp. (Mitchell grass)	Moderate: *Bauhinia, Acacia suberosa, A. farnesiana, Carissa lanceolata, Atalaya hemiglauca, Dolichandrone heterophylla*	Capable of fattening stock. High stocking rates. Inaccessible during wet season
Chrysopogon spp.–*Dichanthium fecundum*	Moderate: *Bauhinia, Acacia suberosa, A. farnesiana, Carissa lanceolata, Atalaya hemiglauca, Dolichandrone heterophylla*	Highly regarded. Subjected to heavy grazing. Large areas scalded and degraded. Parts inaccessible during wet season
Chrysopogon spp.	On frontage similar to above. Elsewhere *Grevillea striata, Bauhinia cunninghamii, Melaleuca* sp.	Moderately high stable stocking rate. High proportion of palatable components
Frontage grass	Good	High fodder production, palatable and nutritious. High stocking rate. Susceptible to degradation under heavy grazing
Fringing grass	Variable	Unattractive to stock
Sehima nervosum–Sorghum spp.	Sparse: *Cochlospermum fraseri, Dolichandrone heterophylla, Atalaya hemiglauca*	Generally unattractive to stock, but utilized in association with better pastures
Sehima nervosum–Dichanthium fecundum	Sparse: *Cochlospermum fraseri, Dolichandrone heterophylla, Atalaya hemiglauca*	Moderate carrying capacity
Plectrachne pungens	Negligible	Low but stable carrying capacity. Drought reserve
Triodia pungens	Moderate: *Bauhinia cunninghamii, Grevillea striata, Atalaya hemiglauca, Carissa lanceolata*	Good subsistence fodder, low but stable carrying capacity

TABLE 31 (*continued*)

Pasture Type (Ground-storey Community)	Top Feed	Remarks
Plectrachne pungens–Chrysopogon spp.	Poor to moderate	Low carrying capacity. Good seasonal fodder when adjacent to frontage pastures or waters
Triodia intermedia	Poor	Unpalatable to stock. Very low carrying capacity. Drought reserve
Triodia wiseana	Poor or absent	Low palatability and low nutritive value. Some utilization in association with better pastures
Triodia inutilis	Absent	Little palatable fodder. Extremely low carrying capacity
Plectrachne bynoei	Present, but inaccessible	Inaccessible, extremely low carrying capacity
Enneapogon spp. and other short grasses	Sparse	Palatable and nutritive. Good fodder for some months after rain but degenerates in long dry periods
Aristida hygrometrica	Sparse	Unpalatable except for short periods after rain. Low carrying capacity. Grass seeds may cause sheep mortality
Aristida browniana	Moderate to good: *Bauhinia cunninghamii* and *Ventilago viminalis*	Palatable for some months after rain, but of low value because sporadic in distribution
Sporobolus virginicus	Absent	Capable of fattening stock, readily grazed at all stages. Good carrying capacity
Xerochloa spp.	Sparse	Highly palatable and nutritious for short periods after rain. Fodder production ceases under drought conditions. Susceptible to degradation
Samphire	Absent	Samphire not grazed. Some fodder from associated plants during favourable season
Beach dune	Sparse	Low fodder production, low carrying capacity

Pasture zones within or bordering the transect

A pasture map of Australia (Fig. 53) compiled by C. S. Christian, C. M. Donald and R. A. Perry ([52]) recognizes a number of pasture zones as occurring within or on the borders of the monsoonal transect. Extracts from *The Australian Environment* are given below to supplement the information already provided above for Queensland, Northern Territory and Western Australia. There are passing references to transect genera under certain of the other pasture zones, while on medium-textured soil or on steppe dominated by spinifex (*Triodia* spp.), *Cenchrus ciliaris* and *C. biflorus* have been established.

MOIST TEMPERATE PERENNIAL GRASS

The dry and wet sclerophyll forests, extensive areas of which are likely always to be preserved as a source of timber, have little grazing value in their native state. When cleared, wet sclerophyll forest commonly gives a good grass cover in which *Themeda australis* is dominant or conspicuous, while areas derived from dry sclerophyll usually give a widely spaced cover of tussock species such as *Poa caespitosa* and *Danthonia pallida*, a tall, coarse and atypical member of this genus. Many other native perennial grasses appear as dominants or co-dominants in various parts of the zone, particularly *Eragrostis* spp., *Bothriochloa decipiens*, *Sporobolus*, *Danthonia*, *Aristida* and *Stipa* species. The survival of many of the taller species, particularly *Themeda australis*, is due in considerable measure to grazing by cattle. In the western Victorian part of the zone where sheep are the principal stock, the short *Danthonia* spp. are prominent.

TEMPERATE SHORT GRASS

This zone, derived principally from savannah and eucalypt savannah woodland, carries the densest concentration of sheep in the Commonwealth, with a carrying capacity on the native pastures of ½–1 sheep per acre. Under grazing, the main genera are *Danthonia* (wallaby grasses) and *Stipa* (spear grasses), the former achieving dominance in southern New South Wales and Victoria, the latter in the northern part of the zone. The wallaby grasses are short, fine-leaved perennials, seldom producing

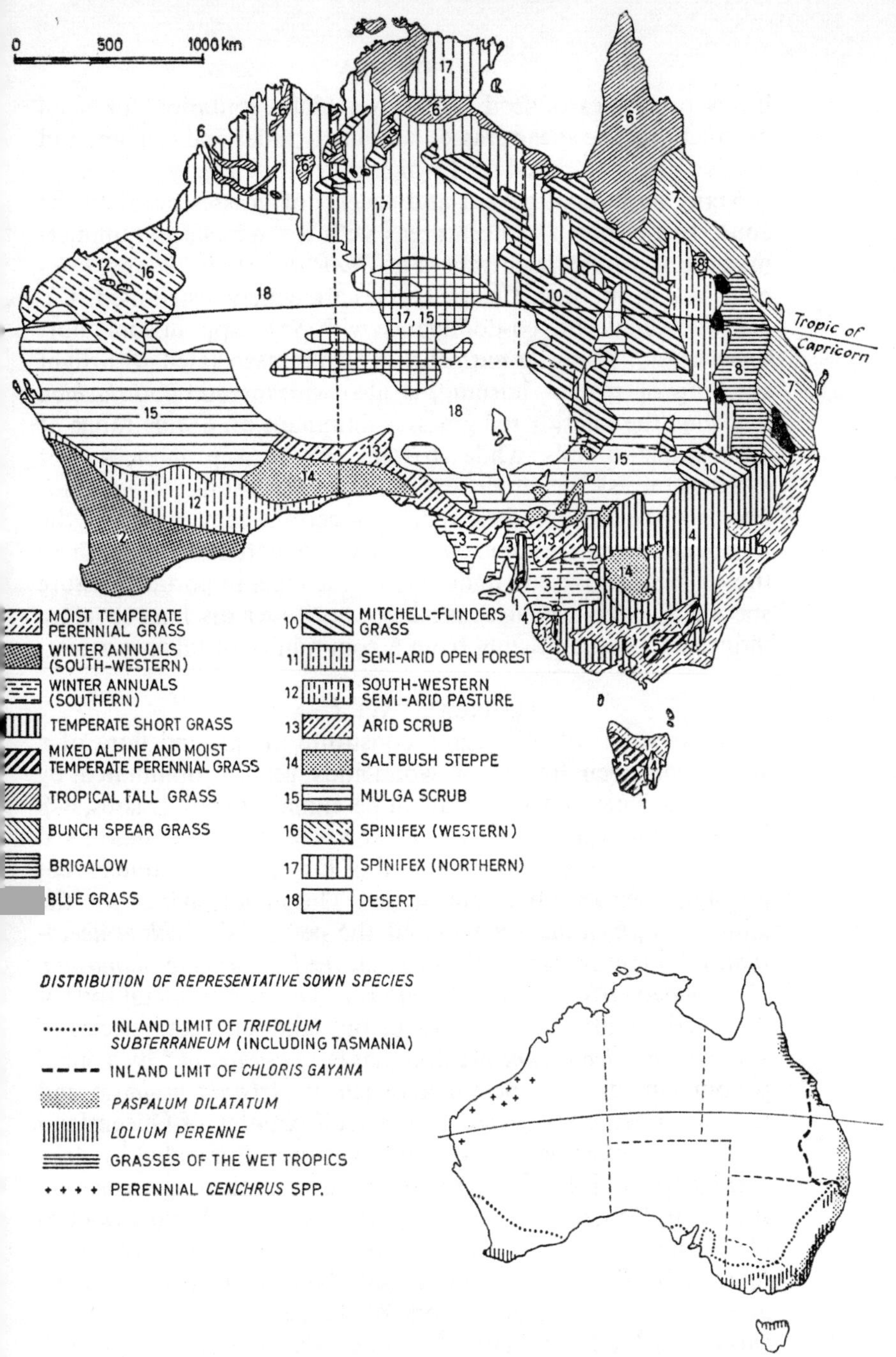

FIGURE 53

Pasture map of Australia, compiled by C. S. Christian, C. M. Donald and R. A. Perry[52]

heavy quantities of feed but with a high reputation for wool production. The spear grasses are mostly taller and coarser, and produce sharp-pointed seed which may cause injury to stock.

Many additional genera and species of grasses occur in the zone. In the north-western areas with a tendency to summer-maximum rainfall such genera as *Panicum*, *Digitaria* and *Eriochloa* are prominent. *Chloris* species are widely distributed, *Ch. acicularis* achieving co-dominance with *Stipa* spp. in some northern areas. *Themeda australis*, generally associated with light stocking or regular burning, is also widespread. *Bothriochloa ambigua*, the inferior red grass, is dominant on a wide range of New England soils, while *Aristida ramosa* is prominent over extensive areas of lighter-textured tableland soils. *Stipa aristiglumis* is the characteristic species of certain heavier soils in the northern part of the zone. Nevertheless, over the zone as a whole the wallaby grasses are undoubtedly the most important pasture species. There is evidence that they constitute a disclimax pasture induced by sheep grazing from a community of taller grasses.

TROPICAL TALL GRASS

The pastures of this region constitute the ground flora of a variety of open forests or woodlands usually dominated by eucalypts. They consist mainly of perennial tussock grasses, 4–8 ft. in height, with some tall annuals which may assume local dominance. These pastures are all of low nutritive value except for short periods after rain or fire. The major species are the annual *Sorghum australiense* and the perennials *Sehima nervosum*, *Chrysopogon* spp., *Themeda australis*, *Sorghum plumosum*, *Heteropogon triticeus* and *H. contortus*. On many skeletal soils of dissected country or remnants of old land surfaces, species of *Triodia* and *Plectrachne* are dominants, accompanied by a small proportion of harsh perennials such as *Aristida pruinosa* and various ephemeral species. In the Gulf country of Queensland there are large areas of low-quality pastures dominated by *Aristida ingrata* and *A. hygrometrica*. In the north the lower areas are subject to flooding. They grade into treeless reed-grass swamps of the heavy-textured estuarine plains or a narrow fringe of *Sporobolus virginicus* around the salt flats. Near the inland margins of the zone the heavy soils of the depressions and valleys carry grassland dominated by *Dichanthium* spp., *Bothriochloa*

various authorities, with classification proposed by M. M. Cole[50]

Oxford Atlas	Library Atlas	Bartholomew's Atlas	Proposed Classification
by D. L. Linton	After Engler, Pole-Evans, Schimper, Shantz and others		
Semi-deciduous tropical forests	Subtropical or tropical grassland overprinted with mallee in the Kimberleys and brigalow in central and eastern Queensland	Open grass woodlands with some cultivation	
	Tropical savannah (grassland with scattered trees and shrubs—the Queensland bush—low eucalypts and brigalow scrub)		
	Subtropical and temperate woodland (eucalypts, brigalow scrub)	Brigalow	Savannah woodland
	Tropical rain forest		
	Eastern and subtropical and temperate rain forest		
Tall grass savannah	Subtropical and temperate woodland (eucalypts, brigalow scrub)	Brigalow	part savannah woodland
		Open grass woodlands with some cultivation	part savannah parkland
	Temperate grassland	Open seasonal grassland	part savannah grassland
		Agriculture	
		Mallee scrub	
Short grass grasslands	Tropical and subtropical grassland	Open seasonal grassland	Savannah grassland
Desert grass savannah	Tropical and subtropical grassland	Open grass woodlands with some cultivation	
	Dry semi-desert (mulga and other scrub) overprinted with mulga and brigalow symbols	Semi-desert Acacia and mixed scrub	Low tree and shrub savannah
	Dry semi-desert (sand, bare rock, spinifex scrub)	Sandy and stony desert	
	Temperate grassland overprinted with mallee and brigalow symbol		

spp., *Eulalia fulva*, *Chrysopogon* spp., and sometimes *Astrebla* spp. The zone is mainly occupied by large cattle stations up to several thousand square miles in extent. Stock-carrying capacity is low, and except on the better pastures of the small flats rarely exceeds five beasts to the square mile. *Imperata cylindrica* var. *major* is an important weed in the small areas of cleared rain forest.

BUNCH SPEAR GRASS

This open forest zone is continuous with the tall tropical pasture zone to the north. The floristic composition, however, differs. The dominant perennial grasses of upland areas in this zone are *Heteropogon contortus* (bunch spear grass) and *Themeda australis* with *Bothriochloa ewartiana* (desert blue grass, desert or forest Mitchell) in the north and *B. decipiens* and *B. intermedia* in the south. These are all grasses of medium height forming a rather open stand of small to medium tussocks. *Heteropogon contortus* appears to increase following ring-barking, grazing and burning, and is the dominant grass over large areas. Natural legumes form only a small proportion of the pasturage, which during the dry season is notably low in protein. An introduced annual legume, *Stylosanthes sundaica*, has shown some promise as a sown species within this zone and has spread naturally over parts of it. The introduced tropical grasses, *Panicum maximum*, *Brachiaria mutica* and *Melinis minutiflora*, are sown in a narrow strip of high rainfall along the Queensland coast. *Paspalum dilatatum*, *Pennisetum clandestinum* and to a less degree lucerne (*Medicago sativa*) are used on the Atherton Tableland, an elevated dairying district west of Cairns.

The better sections of the zone are used for dairying or mixed farming, the remainder for cattle-breeding and fattening.

BRIGALOW

The term, scrub, in most ecological literature is used to describe communities of densely packed bushes. Brigalow scrubs could be variously classified as grassy forests, layered or low-layered forests, low-layered woodlands, savannah woodlands, or even tree savannahs, with brigalow (*Acacia harpophylla*) as the dominant and most conspicuous plant in all communities. There are three primary groupings based on the growth form of the majority of the brigalow plants:

Virgin brigalow communities

Associated grasses: *Paspalidium caespitosum* (brigalow grass), *Chloris* spp. (chiefly *Ch. acicularis*), *Leptochloa* spp. and *Sporobolus* spp.;

Brigalow softwood community

Associated with *Cyperus gracilis*, together with *Ancistrachne uncinatum* and *Chloris unispecia*;

Brigalow in patchy plain community

Predominant grass: *Dichanthium sericeum*.

Cleared brigalow scrub may be sown to Rhodes grass (*Chloris gayana*), otherwise there is a tendency towards the formation of a grassland dominated by *Dichanthium sericeum*.

BLUE GRASS

Reference has already been made to Blake[12] above. Geographically associated with the brigalow zone, and at the higher rainfall limits of the *Astrebla* zone, these areas of tufted grasslands on heavy soils are dominated by species of *Dichanthium*, *Eriochloa*, *Paspalidium* and *Panicum*, with scattered *Astrebla* and *Iseilema* in the north and *Danthonia* and *Stipa* in the south. These are good-quality cattle pastures. In some localities lucerne and annual fodder species are used under cultivation.

MITCHELL-FLINDERS GRASS

The dominant pasture type within this zone is an open tussock grassland of perennial species growing from 1–3 ft. in height, with a lower but prominent annual component occupying the interspaces after rains. The major perennial species are *Astrebla pectinata*, *A. lappacea*, *A. elymoides*, *A. squarrosa* (Mitchell grasses), *Panicum decompositum* and *Eulalia fulva* with *Eragrostis setifolia* in the south and *Chrysopogon fallax* in parts of the north. Of the annual species *Iseilema* spp. (Flinders grasses), *Dactyloctenium radulans* (button grass) and *Panicum whitei* (pepper grass) are common throughout. In addition, there are numerous dicotyledonous herbs, which, in the southern sections, include many winter-growing species. Following rains these pastures make rapid growth and a whole season's grazing may be the product of one fall of rain. This is particularly true in the north. In the south, a good growth of winter annuals may follow the erratic winter rains. Sheep at 4–10 acres per animal are the main stock grazed in the Queensland portion while only cattle at 8–20 beasts per square mile are grazed on these pastures in the Northern Territory.

In the central west of Queensland gidgee, a low tree, is important. In these areas there is a complex mixture of grasses and forbs including *Chloris*, *Sporobolus*, *Paspalidium*, *Enneapogon*, *Digitaria* and *Aristida* spp. Associated pasture communities of importance are those of minor channels and depressions which carry *Chenopodium auricomum*, *Muehlenbeckia cunninghamii*, *Cyperus* spp. and annual forbs and grasses, and those of gravelly rises usually dominated by short and annual grasses, chiefly *Enneapogon* spp., sometimes *Triodia* spp. The *Astrebla* grasslands are mainly treeless but the associated communities include low trees and shrubs, some of which provide top feed. *Acacia farnesiana* is a notable example.

SEMI-ARID OPEN FOREST

This is a mixed zone mainly of eucalypt open forest wherein differing species dominate in different regions. *Eucalyptus populnea* (poplar box) is common and *E. papuana* and *E. melanophloia* are found in the northern part of the zone. The pastures are sparse and include *Aristida*, *Chrysopogon*, *Neurachne*, *Themeda*, *Enneapogon*, *Bothriochloa*, *Eragrostis* and numerous annual grasses and forbs. Spinifex (*Triodia* spp.) occupies much of the northern section. Elements of the Mitchell-Flinders and brigalow zones are included. Top-feed species are important. The zone is used for both cattle and sheep grazing.

Anthropogenic factors

The influence of factors introduced by man, and particularly burning, has been discussed by S. T. Blake (1964)*, with reference especially to the eucalypt-dominant communities that cover such a large part of northern and eastern Queensland. These normally have a good ground cover of grasses of a wide variety of genera and species, chiefly of the Andropogoneae, Paniceae, Eragrosteae, Chlorideae, and the genus *Aristida*. Towards the border of New South Wales, chiefly on the higher hillsides, and becoming more important in New South Wales, *Poa*, *Stipa* and *Danthonia* tend to replace the other genera except *Themeda*. In fact the

* Personal communication.

communities of central coastal Queensland composed of *Themeda/Heteropogon* plus many other genera become largely *Themeda/Poa* on the New South Wales border, with *Danthonia* and *Stipa* becoming prominent farther south.

Removal of the forest cover encourages a denser grass cover but there is little other change. Seasonal burning also seems to have little effect, but severe and repeated burning is associated with local abundance of bracken (*Pteridium*) and *Imperata*. Heavy stocking may lead to the replacement of *Themeda* by *Bothriochloa decipiens* (*B. ambigua* in the cooler parts of New South Wales) and in severe cases of overgrazing the *Bothriochloa* may disappear, leaving (in Queensland) species of *Aristida* abundant. On very sandy soil in the drier parts, *Aristida* spp. may always be very prominent. In south-eastern Queensland there is a strong tendency for the naturalized *Digitaria didactyla* (blue couch) to become very common following any interference with the indigenous grasses.

Towards the north, especially where the rainfall tends to become more or less restricted to the summer, a large number of annual species are found with the perennials. There is little overgrazing, and beyond annual fires (some areas may escape for two or three successive years) man and his animals have not modified the communities to any extent, at least not in historic times.

There was no agriculture and no destruction of forests by man prior to the arrival of the Europeans (unless by casual Malay visitors to the extreme northern coast). The Australian aboriginal used fire as a hunting tool. There is no general agreement as to whether fires are more frequent and more severe since the coming of the European than before, and perhaps any generalization is wrong. The aborigine probably was very intelligent in the use of fire and in many, perhaps most parts of the *Eucalyptus*-dominant areas the white man's fires have damaged the vegetation more than the black man's. This may not apply to the true grasslands—the blue grass (*Dichanthium*) and Mitchell grass (*Astrebla*) communities. Since the introduction of sheep and cattle with the first European settlers, fires have been regarded as major disasters and extensive efforts have been made to reduce the risk of fire and to confine fire to limited areas should one break out, but in spite of all care, fires do occasionally occur. The same is true in the more arid eucalypt-dominant areas.

APPENDIX 1

The occurrence of a monsoonal climate, true or false, and its associated types of vegetation in Africa as proposed by Pédelaborde ([166]) and supported in earlier sections of this book, is perhaps not fully recognized by plant geographers generally. It is therefore timely that the Report of the FAO/UNESCO/WMO Interagency Project on Agroclimatology relating to a semi-arid area, with monsoonal summer rains, in Africa south of the Sahara by J. Cochemé and P. Franquin, has been published ([267]) following the earlier statement to the Reading Symposium (see p. 74).

Cochemé and Franquin note that probably the most permanent and static features in the general circulation affecting the area in West Africa which they have surveyed (see Fig. 14) and, for that matter, the whole globe, are the tropical anticyclones which form two belts encircling the earth on either side of the Equator. These anticyclonic belts are not continuous, but are broken up into individual anticyclones which have customary positions; there is a large intense one usually centred over the Azores, and there are others in West Africa, as shown in Figs. 55a and 55b. The direction in which the anticyclones rotate is determined by a general law; all fluids in motion along the surface of the earth tend to be deflected by the earth's rotation—to the right of the direction of motion in the northern hemisphere and to the left in the southern hemisphere. Consequently the wind circulation around the anticyclones of the northern hemisphere is clockwise, while in the southern hemisphere it is anticlockwise (trade winds). On the Equator side of both anticyclonic belts, winds will therefore blow from east to west.

Thus the trade winds, issuing from the high pressure zones of the two hemispheres, tend to converge towards each other. The intertropical zone (ITCZ) where this convergence takes place receives much heat from the sun and the air begins to rise as it converges and grows warmer. As it rises over the ITCZ, de-

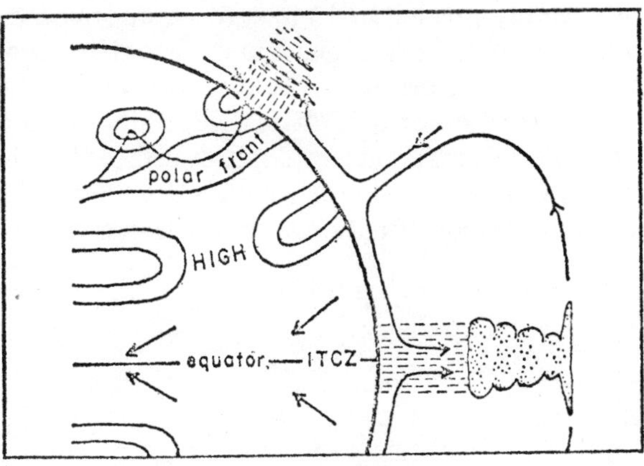

FIGURE 54

General air circulation, showing a cross-section of the atmosphere to the right and the surface of the earth to the left. Descending arrows, surface isobars and 'HIGH' mark the position of the northern belt of anticyclones. Ascent of air and resultant weather are shown over the ITCZ. Arrows converging towards the Equator represent the trade winds. (From Cochemé and Franquin [267])

creasing pressure causes it to expand and grow cooler, its moisture condenses, clouds are formed and rain begins to fall. Freed of its moisture and heated by the release of its latent heat of condensation, this current of rising air is eventually deflected towards the Poles. Part of it feeds the descending air masses over the anticyclonic belt, which are relatively dry and hot. Thus we have complete circulation cells in the vertical plane, resulting in zones of different weather at the surface; the warm and moist ITCZ, and the hot and dry anticyclones.

The ITCZ and the tropical anticyclones move north and south with the seasons. Displacement of the ITCZ over the area studied tends to follow the zenithal position of the sun with a time lag of 4–6 weeks (Fig. 20).

In the summer the relatively low pressure of the ITCZ is sufficiently far north of the Equator to develop anti-clockwise circulation. The oceanic trade winds of the southern hemisphere cross the Equator, moving towards this low pressure. They are laden with readily precipitated moisture, evaporated from the ocean, and are commonly called monsoons. The monsoon is opposed by the dry north-east trade winds, issuing from the anticyclones over the north of Africa, generally called the harmattan. A typical summer meteorological surface chart is shown in Fig. 55a. In the winter, the harmattan dominates the whole of the area and a typical situation at that season is shown in 55b.

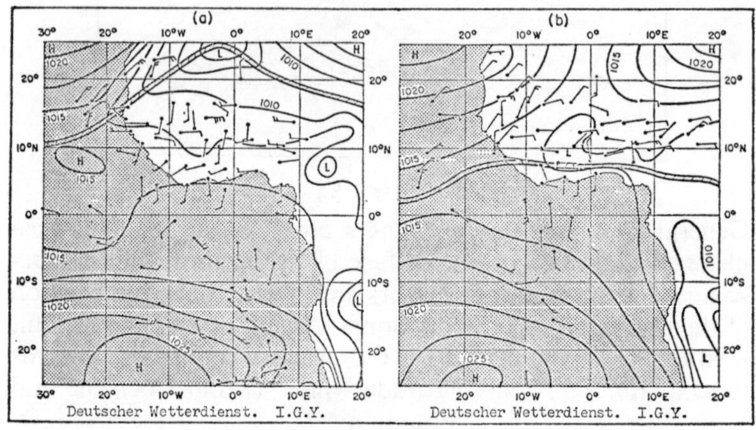

FIGURE 55

West Africa

Typical summer meteorological surface chart, showing monsoon crossing the Equator (a), and typical winter situation, showing the harmattan blowing across the area (b). Surface charts for 1200 GMT (a) 11th August 1958; and (b) 11th December 1957. H, anticyclone; and L, low pressure. The double line shows the position of the ITF (intertropical front). Lines of equal pressure in millibars are shown. Arrows indicate wind direction and the number of barbs, wind strength. (From Cochemé and Franquin[267])

The amount of solar energy incident outside the atmosphere over the area surveyed is high throughout the year, although showing some winter reduction in the north. This, together with the features of atmospheric circulation already described, results in consistently warm climate, dry in winter and with summer rains. It is a clearly recognizable type which can be described as tropical, semi-arid with summer rains, and is defined in the various systems of classification evolved by geographers (Köppen, Thornthwaite and Landsberg). Aubreville defined the climatic zones as reflected in the vegetation south of the Sahara (see Table 6).

In analysing their data, Cochemé and Franquin consider that the very typical monsoonal summer rainfall of the area surveyed shows fairly uniform characteristics. These are sufficiently repetitive to be consistent with the accuracy needed for application to agricultural problems when the rainfall requirements are sufficiently well-defined. Crops in the area, and more particularly those which are indigenous to it, show a multitude of forms adapted severally to the whole range of these climatic conditions. A comprehensive agroclimatic analysis must divide these crops into ecotypes with a narrow range of climatic requirements, which can then be defined with sufficient accuracy for the needs of planning (see pp. 94–6 and Whyte[270]).

As already indicated in the section on West Africa in this book, and in Table 6 and Fig. 2, the general distribution of the vegetation of the area in its present state is zonal. In their section on vegetation, however, Cochemé and Franquin do not refer to any of the species of grasses which in this book are taken as indicators of the monsoonal ecoclimate. Under the heading *Seeded Fallows*, they refer to *Andropogon gayanus* as being the best local species available for seeding. This species is found naturally at least as far north as the 400 mm. isohyet where the dry season lasts about 200 days.

Certain conclusions are made regarding homologous areas, where the basic tropical circulation described above has resulted in semi-arid climates and agricultural utilization of the land comparable to those of the area surveyed. Nowhere, however, is there such a large expanse, according to Cochemé and Franquin, free from the influence of high ground or sea. The two most homologous areas recognized by these writers are north-west India and the adjacent part of West Pakistan in the northern hemisphere, and the north of Australia in the southern hemisphere. Neither, according to Cochemé and Franquin, presents quite the same climatic latitudinal zones as the African area, and also, for reasons other than climate, their exploitation has so far been very different.

As far as Africa itself is concerned, the conditions found in the area surveyed by the FAO/UNESCO/WMO Interagency Project extend beyond the border of Chad into the Sudan, and it is only in the extreme east of the latter country that the mountains of Ethiopia and the Red Sea distort the simple tropical atmospheric circulation which prevails to the west. South of the Equator in Africa, Cochemé and Franquin recognize other areas of homologous climate, notably in Malawi, Rhodesia and further west. In North and South America, homologous climates of summer rainfall do occur, but with the possible exception of Brazil, only in rather limited areas (see p. 27). This Cochemé and Franquin attribute to the influence of both high ground and the seas.

It will be seen, in comparing the results of this U.N. Interagency Survey with the approach adopted in the main text of this book, that the Interagency Survey is primarily a meteorological study. It does not, in the opinion of plant geographers, give enough attention to the zonal distribution of plant communities, and the genera and species of grasses in particular which are found in these plant communities. References to physiognomic types such as wooded and grass steppe, woods and savannah, and to areas of short or taller grasses are not sufficient as plant indicators of the monsoonal environment. The zones of homologous climates may be correct in the rather narrow agroclimatological sphere, as indicating specific sub-types within the total transect of comparable monsoonal climates.

However, it is considered that the repeated occurrence of characteristic genera and species of grasses throughout this transect of tropical grasslands is sufficient justification for the mapping of the latitudinal zone north and south of the Equator (Fig. 1) in which the dominant climatic factor is the seasonality of summer rainfall. It is much easier for taxonomic geographers and historians to accept continuous distribution of certain genera along an ecoclimatic zone between West Africa and Rajasthan and of the same or other genera between India and northern Australia, than to accept discontinuous distribution between three 'homologous' areas so far apart from one another.

The basic ecoclimatic pattern along the transect may be combined with aridity in Africa, Arabia and north-west India and

Pakistan, with average to extreme degrees of coldness in its north-east limits in China and Japan, passing through the equatorial ever-wet climates in the islands of south-east Asia to a full recurrence of monsoonal conditions in northern Australia. It is considered that the suggestion made in the Interagency Survey that a monsoonal climate occurs south of the Equator in Malawi and Rhodesia is open to considerable debate.

It is perhaps the following conclusion reached by Cochemé and Franquin that provides the best support for the adoption of the monsoon as the basis for the recognition, classification and mapping of a major ecoclimatic zone: 'Consequently it is, on the whole, the actual duration of the rains rather than the adequacy of the water supply during the rains which is the most important agroclimatic limitation in the Area. In order to define more specifically this duration in terms of agronomic productivity potentials, "availability-of-water" periods were used, based on a comparison of rainfall and evapotranspiration. The length of the "humidity" period, during which rainfall exceeds potential evapotranspiration, varies with latitude fairly uniformly, increasing from north to south from nil to 140 days.'

APPENDIX 2

Detailed legends of plates 1–13: Vegetation of Mauritania and adjacent areas in Mali. (*Photos Ch. Rossetti*)

1–9: Rainfall gradient on sandy soils

1. Site 61/88. (Site numbers refer to the Report: *Observations sur les sols et la végétation en Mauritanie du SE et sur le bordure adjacente du Mali* by P. Audry and Ch. Rossetti.[268]
   Approximate position 15° 50′ N/9° 25′ W, Western Hodh. Annual rainfall (summer, one season only) about 500 mm. Savannah with *Andropogon gayanus* and *Hyparrhenia dissoluta* and scattered small trees (*Combretum glutinosum*, *Sclerocarya birrea*). These physiognomic types which contain Sudanese floristic components represent intrusions of formations belonging to more humid regions. They are often localized on flattened dunes. A red-brown soil has been identified in this sand. The annual production from the herbaceous layer may be estimated at 2,500 kg. of dry matter per hectare.
2. Some kilometres from the site illustrated in Pl. 1, in the same edaphic conditions; a formation resulting from the destruction of the former savannah by man for cultivation (here of *Arachis hypogaea*). The tall perennial grasses are replaced by an annual sward based on *Cenchrus biflorus*, the trees disappear and are replaced by shrubs of the same or different species (*Guiera senegalensis*, *Boscia*, etc.).
3. Same site as Pl. 2. The physiognomic contrast with the savannah is of edaphic nature. The brown soil is a loamy sand. There is only a thin cover of annuals (*Aristida funiculata*, *Schoenefeldia gracilis*, etc.); patches of *Cymbopogon schoenanthus* and *C. giganteus* (perennials) indicating increased moisture in the depressions; scattered shrubs of

Boscia senegalensis and *Acacia seyal.* In the foreground, a plant of *Combretum glutinosum.* Annual production from the herbaceous vegetation (annuals) is less than 1,000 kg. of dry matter per hectare.

4. Position 15° 47′ N/11° 37′ W. Under the same rainfall conditions as the sites illustrated in Plates 1 to 3, but on possibly more compact sand, there is a characteristic physiognomic variant: an annual sward with a short herbaceous stratum (*Cenchrus biflorus*, *Blepharis* sp., *Tribulus* sp., *Gisekia* sp.), and a shrub and tree storey of a density of about 100 plants per hectare. The photograph shows fine specimens of *Sclerocarya birrea* (left), one *Balanites aegyptiaca*, sometimes accompanied by *Sterculia setigera.* It is of course possible that the herbaceous layer is the result of the elimination of a savannah storey similar to that in Pl. 1. Certain partial effects of closure would in fact seem to indicate that the regeneration of this savannah vegetation is extremely slow, or even non-existent under certain conditions of competition by annuals.
5. Position 16° 20′ N/9° 22′ W; Msilé plain; estimated annual rainfall: 350 mm.; Site No. 61/77. Brown soil developed from sand. Short-duration annual sward (*Cenchrus biflorus*, *C. prieurii*, *Mollugo* sp., *Aristida stipoides*, etc.) with small trees and shrubs (*Acacia raddiana* and *Balanites aegyptiaca*).
6. The same community in a site near that in Pl. 5, after obvious biotic interference. The tree storey has disappeared almost entirely. Steppe elements belonging to more northern regions (*Aristida pallida*, *Panicum turgidum*, etc.) have established themselves as a consequence of probable wind disturbance in the plain, due to denudation.
7. The belt of annual grassland, as shown in Plates 4 to 6, extends with variations in botanical composition over a rainfall zone between the 400- and 250-mm. isohyets. Near the latter isohyet, steppe grassland appears, which may be defined as an annual grass cover interspersed with patches of narrow-leaved perennials, less than 80 cm. high, often localized on areas of sandy soil where wind erosion may be clearly observed. These steppe grasslands form the transition between the annual grasslands to the south and the continuous steppe cover to the north. In the region of Tim-

bédra (16° 25′ N/9° 02′ W; annual rainfall about 260 mm.), on coarse-textured sand, was a good example of such a steppe grassland, here with scattered small trees and shrubs (*Combretum aculeatum*, *Grewia bicolor*, *Boscia senegalensis* and others), and with a herbaceous layer composed primarily of annuals (*Aristida mutabilis*, *Cenchrus biflorus*, etc.), together with patches of steppe elements (*Aristida pallida* and *Cymbopogon schoenanthus*; the latter species shows no rainfall differentiation).

8. Dahr of Oualata, position 17° 17′ N/6° 58′ W; annual rainfall 100 mm.; sandy plateau; Site No. 61/36. Steppe of *Panicum turgidum* with annuals (*Aristida stipoides*, *Gynandropsis* sp., *Heliotropium* sp., *Latipes* sp., *Sesamum* sp., etc.) and a tree and shrub storey (*Acacia ehrenbergiana*, *Boscia senegalensis*, *Commiphora africana*, *Maerua crassifolia*). This is the northern limit of distribution for most of the woody species and the annuals. These *Panicum turgidum* steppes cover a considerable area in north-western Africa; they are at least 1° Lat. wide. Their southern limit is not precisely governed by rainfall, but rather by the topography of the sand. The more dissected it is, the further south these steppes may be found.
9. Further north, *Panicum turgidum* is replaced by *Aristida pungens*; the woody vegetation becomes thinner (disappearance of certain species and reduction in the density of the remaining ones), and shrubby species such as *Calligonum comosum* and, even further north, *Cornulaca monacantha* appear.

   Here, in the Aouker, with strongly undulating sands in 'aklé' dune pattern, is an almost pure steppe of *Aristida pungens* with a very scattered woody vegetation (*Calligonum*, *Leptadenia*, *Acacia raddiana*). ('Aklé' is the Mauritanian local name for a dune massif with overlapping dunes composed of usually very unstable sand.) The annual rainfall is still of the order of 90 mm. (approximate position 18° 05′ N/9° 30′ W). These communities become more open as one proceeds further north; on these sand dunes *Aristida pungens* is found practically alone; it extends almost to the extreme vegetation limit in the 'desert', where sometimes only *Cornulaca* is found.

10–13: Edaphic communities in the region with less than 150 mm. rainfall

10. Halophilous communities cover only limited areas in the whole West African continental Sahara, as compared with other continental deserts, and especially with the northern Sahara. These communities are found only in small closed basins with no drainage outlet. They are far more extensive in the Atlantic littoral from the 16th parallel N to Morocco, but are limited to a coastal strip varying in width from only a few hundred metres to some 15 km.

In south-western Mauritania, some halophilous communities are found on the edge of the Primary sandstone escarpment between Tagouraret (17° 30′ N/7° 30′ W) and Tichit (18° 20′ N/9° 30′ W). The site shown (in the region of Oujaf, 17° 46′ N/7° 52′ W; rainfall 80 mm.) is a saltpan (*sebkha*) which is completely bare in the centre and has a shrubby fringe of *Salvadora persica* (left) and *Suaeda fruticosa* (right). On the non-saline sand in the foreground are some plants of *Panicum turgidum.*

11, 12 & 13 Sites No. 59/75 and following; reference position 18° 20′ N/1° 20′ E; region of Kidal (Adrar of the Iforas, north-eastern Mali); annual rainfall 130 mm. The characteristics of the gradient illustrated for Mauritania were also found, with small variations, in 1959 in eastern Mali. Plates 11 to 13 show the characteristic physiognomic gradient in a wadi in the Adrar des Iforas. In dissected terrain, Saharan wadis show three different plant habitats: the tributary gullies; one or more batha (beds) into which the gullies drain; and a flood plain where the water infiltrates rapidly because of its loss of velocity and where there is thus usually no surface flow. These three situations are physiognomically very different.

Species with a more southerly distribution often survive in the gullies. Thus Pl. 11 shows thickets of *Balanites aegyptiaca*, *Grewia tenax* and *Salvadora persica*, separated by areas of *Andropogon gayanus*, a savannah grass which is here associated with *Dichanthium annulatum*, *Cymbopogon schoenanthus* and others.

As the gullies widen and the now more accessible sites become drier due to grazing and cutting, these floristic elements disappear. Along the wadi bed there are nearly always fairly luxuriant riparian communities. Pl. 12 shows dense thickets of *Balanites aegyptiaca*, *Combretum aculeatum* and *Acacia scorpioides*, interrupted by dense stands of *Panicum turgidum* and other species. Away from the banks, on the compact sand of the higher ground is a poor annual grassland of *Aristida mutabilis*, associated with *Acacia raddiana*.

The batha is sometimes interrupted by intermediate flats where the stream bed disappears. In the terminal plain, with its rather deep deposits, there is a patchwork vegetation determined by the degree of wind disturbance. Fig. 13 shows a short-duration annual grassland with small trees on sandy clay soil (loam + clay more than 15 per cent). Note the complete cover of annual grasses with a short growth cycle (*Aristida adscensionis*, *Schoenefeldia gracilis*) with herbs such as *Cassia tora*. The shrubs and trees are *Acacia ehrenbergiana* and *A. raddiana*.

BIBLIOGRAPHY

1. Administration of Territory of Papua and New Guinea and Unesco Science Cooperation Office for South East Asia (1960). *Symposium on the Impact of Man on Humid Tropics Vegetation*, 402 pp.
2. Albertson, F. W., Romanek, G. W. and Riegel, A. (1957). Ecology of drought cycles and grazing intensity on grasslands of Central Great Plains, *Ecol. Monogr.*, 27: 27–44.
3. —— and Weaver, J. E. (1945). Injury and death or recovery of trees in prairie climate, *Ecol. Monogr.*, 15: 393–433.
4. Association pour L'etude Taxonomique de la Flore d'Afrique Tropicale (1959). *Vegetation Map of Africa south of the Tropic of Cancer*, Explanatory notes by R. W. J. Keay. Published with assistance of UNESCO (second edition in preparation). Oxford Uni. Press, London, 24 pp. and 1 map.
5. Aubreville, A. (1949). *Climats, Forêts et Désertification de l' Afrique tropicale.* Paris, Soc. d'Editions géographiques, maritimes et coloniales, 352 pp.
6. Bagnouls, F. and Meher-Homji, V. M. (1959). Types bioclimatiques du Sud-Est Asiatique, *Inst. fr. Pondichéry Tr. Sect. Sci. Tech.*, I, 207–46.
7. Beadle, N. C. W. and Costin, A. B. (1952). Ecological classification and nomenclature, *Proc. Linn. Soc. NSW*, 77: 61–82.
8. Beaujeu-Garnier, J. (1965). *Trois Milliards d'Hommes*, Hachette, Paris. 402 pp.
9. Bennett, Erna (1965). Plant Introduction and Genetic Conservation: Genecological Aspects of an Urgent World Problem, *Scott. Plant Breeding Station Record*, 27–113.
10. Bews, J. W. (1929). *The World's Grasses: Their Differentiation, Distribution, Economics and Ecology.* Longmans, Green and Co., London, New York, Toronto, 408 pp.
11. Bharadwaj, O. P. (1961). The Arid Zone of India and Pakistan. In A History of Land Use in Arid Regions, *Arid Zone Research Series*, XVII, UNESCO, Paris, 143–73.
12. Blake, S. T. (1938). The plant communities of western Queensland and their relationships, with special reference to the grazing industry, *Proc. Roy. Soc. Queensl.*, 49: 156–204.
13. —— (1940). The interrelationships of the plant communities of Queensland, *Proc. Roy. Soc. Queensl.*, 51: 24–31.
14. —— (1953). Botanical contributions to the Northern Australia Regional Survey. I. Studies on Northern Australian species of *Eucalyptus, Aust. J. Bot.*, 1: 185–352.
15. Blumenstock, D. L. (1958). Distribution and characteristics of tropical climates, *Proc. 9th Pacific. Sci. Congr.*, Bangkok, 1957, 20: 3–24.
16. Boaler, S. B. and Hodge, C. A. H. (1964). Observations on vegetation arcs in the northern region, Somali Republic, *J. Ecol.*, 52: 511–44.

17. Bor, N. L. (1938). The vegetation of the Nilgiris, *Indian For.*, 64: 600–9.
18. —— (1942). Ecology: Theory and Practice. Presidential address to the Botany Section, *Proc. 29th Indian Sci. Congr., Baroda* Part 2, 145–79.
19. —— (1960). *Grasses of Burma, Ceylon, India and Pakistan (except Bambuseae)*. Pergamon Press, Oxford, London, New York, Paris, 767 pp.
20. Brennan, J. P. M. (1965). The geographical relationships of the genera of Leguminosae in Tropical Africa, *Webbia*, 19: 545–78.
21. Buchanan, K. (1966). *The Chinese People and the Chinese Earth.* G. Bell and Sons Ltd., London, 94 pp.
22. Budowski, G. (1968). *Climatological Data and Natural Vegetation*, Paper to UNESCO Symposium on Methods in Agroclimatology. Reading, 23–30 July 1966. In Agroclimatological Methods. *Natural Resources Research Series*, VII. UNESCO, Paris.
23. Budyko, M. I. (1968). *Solar Radiation and the use of it by Plants*, Paper to UNESCO Symposium on Methods in Agroclimatology. Reading, 23rd–30th July 1966. In Agroclimatological Methods. *Natural Resources Research Series*, VII. UNESCO, Paris.
24. Burbidge, Nancy T. (1960). The phytogeography of the Australian region, *Aust. J. Bot.*, 8: 75–211.
25. Butzer, K. W. (1965). *Environment and Archaeology. An Introduction to Pleistocene Geography*. Methuen and Co., London, 524 pp.
26. Cain, S. A. (1944). *Foundations of Plant Geography*. Harper and Brothers, New York and London, xiv + 556 pp.
27. —— (1950). Life-forms and phytoclimate, *Bot. Rev.*, 16: 1–32.
28. Capot-Rey, R. (1953). *Le Sahara français*. Presses universitaires de France, Paris, 564 pp.
29. Carvalho, G. and Gillet, H. (1960). *Catalogue raisonné et commenté des plantes de l'Ennedi (Tchad septentrional)*. Office Anti-acridien, Alger. 158 pp.
30. Champion, H. G. (1936). A Preliminary Survey of the Forest Types of India and Burma, *Indian For. Rec. (N/S) Silv.*, 1–286 + i–viii.
31. —— and Seth, S. K. (1968). *A Revised Survey of Forest Types of India*. Manager of Publications, Delhi.
32. Chapman, R. F. (1957). Observations on the feeding of adults of the Red Locust (*Nomadacris septemfasciata* (Serville)), *Brit. J. Anim. Behav*, 5: 60–75.
33. —— (1959). Field observations on the behaviour of hoppers of the Red Locust, *Anti-Locust Bulletin*, 33: 46–7.
34. Chaudhri, I. I. (1960). The vegetation of the Kaghan Valley, *Pak. J. For.*, 10: 285–94.
35. Chevalier, A. (1924). Le role joué par les migrations humaines dans la répartition actuelle de quelques vegétaux, *Congrès Assoc. Franc. Avan. Sci.* Liege. 990–6.
36. —— (1928). La végétation montagnarde de l'Ouest Afrique et sa genèse, *Mém. Soc. Biogéographie*, 2: 221–9.
37. —— (1938). *Flore vivante de l'Afrique occidentale française*. Musée d'histoire naturelle, 57, rue Cuvier, Paris 5ème. Introduction, subdivision chronologique, pp. v–xxxi.
38. Chinese People's Republic, Ministry of Agriculture (1959). *Illustrated Manual of Forage Plants of China*. (English translation published by the) Office of Technical Services of the U.S. Dept. of Commerce, Washington, D.C. 25 (1963).
39. Christian, C. S. (1952). Regional land surveys, *J. Aust. Inst. Agric. Sci.*, 18: 140–6.

40. —— (1958). The concept of land units and land systems, *Proc. 9th Pacific Sci. Congr., Bangkok, 1957*, 20: 74–81.
41. —— (1959). The eco-complex in its importance for agricultural assessment. In Biogeography and Ecology in Australia, *Monographiae Biologicae*, 8: 587–605.
42. ——, Norman, M. J. T. and Arndt, W. (1958). Some investigations related to livestock production in Northern Territory of Australia, *Proc. 9th Pacific Sci. Congr., Bangkok, 1957*, 2: pp. 1–6.
43. —— and Perry, R. A. (1953). The systematic description of plant communities by the use of symbols, *J. Ecol.*, 44: 100–5.
44. —— and Shaw, N. H. (1951). Protein status of Australian tropical and sub-tropical pastures. Brit. Commonw. Sci. Offic. Conf., *Spec. Conf. in Agric., Aust. 1949. Proc.* HMSO, London, 225–40.
45. —— and Slatyer, R. O. (1958). Some observations on vegetation changes and water relationships in arid areas. In Climatology and Meteorology: Proceedings of the Canberra Symposium 1956. *Arid Zone Res.*, 11: 156–8.
46. —— and Stewart, G. A. (1952). General Report on Survey of Katherine-Darwin region, 1946, *Land Research Series* 1, CSIRO, 151 pp.
47. ——, —— and Perry, R. A. (1960). Land research in northern Australia, *Aust. Geogr.*, 7: 217–31.
48. Cocheme, J. (1968). *FAO/UNESCO/WMO Agroclimatological Survey of a Semi-Arid Area in West Africa south of the Sahara.* Paper to UNESCO Symposium on Methods in Agroclimatology. Reading, 23rd–30th July 1966, In Agroclimatological Methods. *Natural Resources Research Series*, VII. UNESCO, Paris.
49. Cole, Monica M. (1963). Vegetation and geomorphology in Northern Rhodesia. An aspect of the distribution of the savannah in Central Africa, *Geogr. J.*, 129: 290–305.
50. —— (1963). Vegetation nomenclature and classification with particular reference to the savannahs, *South Afr. Geog. J.*, 45: 3–14.
51. Commonwealth Scientific and Industrial Research Organization (1952). Survey of Barkly Region 1947–8. Extracts from *Land Research Series*, 3: 55 pp.
52. —— (1960). *The Australian Environment* (third edition). Cambridge Uni. Press, London and New York, 151 pp.
53. —— (1964). General Report on Lands of the Leichhardt–Gilbert Area, Queensland, 1953–4, *Land Research Series*, 11: 244 pp.
54. —— (1965). General Report on Lands of the Tipperary Area, Northern Territory, 1961, *Land Research Series*, 13: 112 pp.
55. Cougoulis, J. N. (1965). L'amélioration des principales pistes de bétail de la région est du pays. Rapport au Gouvernement de la République du Tchad, *FAO EPTA Report No. 1959.* Rome, mimeogr., 15 pp.
56. Cressey, G. B. (1963). *Asia's Lands and Peoples* (third edition). McGraw-Hill Inc., New York, 663 pp.
57. Crocker, R. L. (1959). Past climatic fluctuations and their influence upon Australian vegetation. In A. Keast, R. L. Crocker and C. S. Christian (eds.), *Biogeography and Ecology in Australia*, Uitgevereij Dr. W. Junk, Den Haag, 640 pp. (282–90).
58. —— and Wood, J. G. (1947). Some historical influences on the development of the South Australian vegetation communities and their bearings on concepts and classification in ecology, *Trans. Roy. Soc. S. Aust.*, 71: 91–136.
59. Dabadghao, P. M. (1961). Types of grass covers in India and their management, *Proc. 8th Internat. Grassl. Cong. 1960*, 226–30.

60. Dansereau, P. (1957). *Biogeography. An Ecological Perspective.* Ronald Press Co., New York.

61. Davey, P. M. (1954). Quantities of food eaten by the Desert Locust (*Schistocerca gregaria* (Forsk.)) in relation to growth, *Bull. Ent. Res.*, 45: 539–51.

62. De Leeuw, P. N. (1965). The role of savannah in nomadic pastoralism: some observations from western Bornu, Nigeria, *Netherl. J. agric. Sci.*, 13: 178–89.

63. De Winter, B. (1965). The South African Stipeae and Aristideae (Gramineae), *Bothalia*, 8: 199–404.

64. Dobby, E. H. G. (1960). *Southeast Asia.* Uni. London Press, London, 7th ed., 415 pp.

65. —— (1961). *Monsoon Asia.* Uni. London Press, London, 381 pp.

66. Eig, A. (1931–2). Les éléments et les groupes phytogéographiques auxiliaires dans la flore palestinienne. I, II., *Feddes Repert. Spec. Nov. Regni. Veg. Beih.*, 63: 1–201.

67. —— (1938). On the phytogeographical subdivision of Palestine, *Palestine J. Bot. Jer. Ser.*, 1: 4–13.

68. —— (1946). Synopsis of the phytosociological units of Palestine, *Palestine J. Bot. Jer. Ser.*, 3: 183–246.

69. Fosberg, F. R., Garnier, B. J. and Kuchler, A. W. (1961). Delimitation of the humid tropics, *Geogr. Review*, 51: 333–47.

70. Gillett, J. B. (1941). The plant formations of Western British Somaliland and the Harar Province of Abyssinia, *Kew Bulletin*, 2: 37–75.

71. Gilli, A. (1955). Beitraege zur Flora Vorderindiens, *Verhandl. zool.-bot. Ges. Wien*, 95: 168–80.

72. —— (1957). Beitraege zur Flora des Karakorum (Von der oesterr. Karakorum-Expedition 1956 gesammelte Pflanzen), *Oest. bot. Z.*, 104: 303–12.

73. —— (1958). Beitraege zur Flora Afghanistans I, *Feddes Repert.*, 61: 86–92.

74. —— (1962). Beitraege zur Flora Afghanistans II. Monocotyledones, *Feddes Repert.*, 64: 204–31.

75. —— (1962). Botanische Ergebnisse der oesterreichischen Karakorum-Expedition 1961, *Oest. bot. Z.*, 109: 108–12.

76. Good, R. (1964). *The Geography of the Flowering Plants* (third edition). Longmans, Green and Co. Ltd., London, 518 pp.

77. Gupte, S. C., Chinnamani, S. and Rege, N. D. (1965). Personal communication.

78. —— and Rege, N. D. (1965). Improvement of natural grasslands on the Nilgiri Plateau, *Indian For.*, 91: 115–22.

79. Handel-Mazzetti, H. (1930). The phytogeographic structure and affinities of China, *Abstr. 4th Internat. bot. Congr. Cambridge*, 315–19.

80. Hare, F. K. (1953). *The Restless Atmosphere.* Hutchinson, London.

81. Harlan, J. R. and De Wet, J. M. J. (1963). Role of apomixis in the evolution of the *Bothriochloa/Dichanthium* complex, *Crop Sci.*, 3: 314–16.

82. Harrison, M. N. (1955). *Report on a grazing survey of the Sudan.* Mimeogr.

83. Hartley, W. (1950). The global distribution of tribes of the Gramineae in relation to historical and environmental factors, *Aust. J. agric. Res.*, 1: 355–73.

84. —— (1954). The agrostological index. A phytogeographical approach to the problems of pasture plant introduction, *Aust. J. Bot.*, 2: 1–21.

85. Hayman, D. L. (1960). The distribution and cytology of the chromosome
86. races of *Themeda australis* in southern Australia, *Aust. J. Bot.*, 8: 58–68.
Hedberg, O. (1963). Afroalpine flora elements, *Webbia*, 19: 519–29.

87. Heine, H. (1963). Four representatives of the Saharo-Sindian element in the flora of Senegal and Mauritania, *Kew Bull.*, 16: 203–7.

88. Hemsley, W. B. and Pearson, H. H. W. (1902). The flora of Tibet or High Asia, *J. Linn. Soc. Bot.*, 35: 124–265.
89. Hepper, F. N. (1965). Preliminary account of the phytogeographical affinities of the flora of West Tropical Africa, *Webbia*, 19: 593–617.
90. Heyligers, P. C. (1965). Vegetation and Ecology of the Port Moresby–Kairuku Area. In Lands of the Port Moresby–Kairuku Area, Papua–New Guinea, *Land Research Series*, 14: CSIRO, Australia, 182 pp. with map.
91. Holmes, C. H. (1946). Grasslands and their afforestation in Ceylon, *Indian For.*, 72: 6–11.
92. Holttum, R. E. (1940). The uniform climate of Malaya as a barrier to plant migration, *Proc. 6th Pacific Sci. Congr. 1939*, 4: 669–71.
93. —— (1954). *Plant Life in Malaya.* Longmans, Green and Co., London, 254 pp.
94. Hubbard, C. E. (1934). Gramineae. In J. Hutchinson, *The Families of Flowering Plants, II, Monocotyledons.* Macmillan and Co. Ltd., London, 243 pp. (199–299).
95. —— (1948). Gramineae. In J. Hutchinson, *British Flowering Plants.* P. R. Gawthorn Ltd., London, 372 pp. (284–348).
96. —— and Vaughan, R. E. (1940). *The Grasses of Mauritius and Rodriguez.* Crown Agents for the Colonies, London, 128 pp.
97. Hutchinson, J. (1959). *The families of flowering plants.* Clarendon Press: Oxford University Press, 2nd ed., 1: 521 pp.
98. International Union for Conservation of Nature (1964). 9th Technical Meeting (Nairobi, Sept. 1963). *IUCN publications new series*, 4, Part III. The Impact of Man on the Tropical Environment, 210–19.
99. Jacques-Felix, H. (1962). *Les graminées d' Afrique tropicale*, Paris: Inst. Rech. Agron. Trop., 1: 356 pp.
100. Johannesburg University of Witwatersrand (1961). *Climatological Atlas of Africa.* Govt. Printer, Pretoria, acting for Commission for Tech. Coop. in Africa south of Sahara and Scientific Council for Africa south of Sahara. Joint Project No. 1.
101. Johnson, R. W. (1964). *Ecology and Control of Brigalow in Queensland.* Queensland Dept. of Primary Industries, Govt. Printer, Brisbane, 92 pp.
102. Johnston, A. and Hussain, I. (1963). Grass cover types of West Pakistan, *Pakistan J. For.*, 13: 239–47.
103. Kassas, M. (1952). Habitat and plant communities in the Egyptian desert I. Introduction, *J. Ecol.*, 40: 342–51.
104. —— (1953). Habitat and plant communities in the Egyptian desert II. The features of a desert community, *J. Ecol.*, 41: 248–56.
105. —— (1953). Land forms and plant cover in the Egyptian desert, *Bull. Soc. Géogr. d'Egypte*, 26: 193–205.
106. —— (1956). Landforms and plant cover in the Omdurman desert, Sudan, *Bull. Soc. Géogr. d'Egypte*, 29: 43–58.
107. —— (1956). The mist oasis of Erkwit, Sudan, *J. Ecol.*, 44: 180–94.
108. —— (1957). On the ecology of the Red Sea coastal land, *J. Ecol.*, 45: 187–203.
109. —— (1966). *Plant Life in Deserts.* In *Arid Lands.* Methuen, London/ UNESCO, Paris, 145–80.
110. —— and Girgis, W. A. (1965). Habitat and plant communities in the Egyptian Desert VI. The units of a desert ecosystem, *J. ecol.*, 53: 715–28.
111. —— and Imam, M. (1954). Habitat and plant communities in the Egyptian desert III. The wadi bed ecosystem, *J. Ecol.*, 42: 424–41.
112. —— and —— (1957). Climate and microclimate in the Cairo desert, *Bull. Soc. Géogr. d'Egypte*, 30: 25–52.

113. —— and —— (1959). Habitat and plant communities in the Egyptian desert IV. The gravel desert, *J. Ecol.*, 47: 289–310.
114. —— and ZAHRAN, M. A. (1962). Studies on the ecology of the Red Sea coastal land I. The district of Gebel Ataqa and El-Galala, El-Baharaiya, *Bull. Soc. Géogr. d'Egypte*, 35: 129–75.
115. KAWAKITA, J. (1956). Vegetation, Crop Zones. In *Land and Crops of Nepal Himalaya. Scientific Results of Japanese Expedition to Nepal Himalaya 1952–53*, Vol. 2. Kyoto, pp. 1–65 and 67–93.
116. KENDREW, W. G. (1963). *The Climates of the Continents*, 5th edition, O.U.P., 608 pp.
117. KERNICK, M. D. (1966). Report to the Government of Kuwait on plant resources, range ecology and fodder plant introduction. FAO Report No. TA 2181, Rome, 95 pp.
118. KHAN, A. and BHATTI, A. G. (1956). Effect of closure on growth of grasses, *Pak. J. For.*, 6: 187–90.
119. KINGSNORTH, G. W. (1963). *Africa South of the Sahara.* Cambridge Uni. Press, 160 pp.
120. KITAMURA, SIRO (1955). Flowering plants and ferns. In Kihara, H. (ed.), *Fauna and Flora of Nepal Himalaya. Scientific results of the Japanese expedition to Nepal Himalaya 1952–53.* Vol. 1, Kyoto University, pp. 73–290.
121. —— (1960). *Flora of Afghanistan. Results of the Kyoto University Scientific Expedition to the Karakoram and Hindukush 1955.* Vol. II, Kyoto Uni., ix + 486 pp.
122. —— (1964). *Plants of West Pakistan and Afghanistan. Results of the Kyoto University Scientific Expedition to the Karakoram and Hindukush, 1955.* Vol. III, Kyoto Uni., vii + 283 pp.
123. KMOCH, H. G. (1964). L'amélioration des pâturages et de la production fourragère—Report to Government ol Upper Volta. FAO EPTA Report No. 1873, Rome, 40 pp.
124. KØIE, M. and RECHINGER, K. H. (1965). Gramineae, by N. L. Bor and A. Melderis. In *Symbolae Afghanicae* Vol. VI, Biologiske Sknfter Vol. 14: 4. Royal Danish Academy of Sciences and Letters, Copenhagen. 36–94.
125. KOPPEN, W. (1923). *Die Klimate der Erde.* Berlin: Leipzig, Walter de Gruyter, 369 pp.
126. —— (1936). *Das geographische System der Klimate. Handbuch der Klimatologie.* Berlin, Bornträger, 44 pp.
127. LANDSBERG, H. E., LIPPMAN, H., PAFFEN, K. H. and TROLL, C. (1965), *Weltkarten zur Klimakunde/World maps of climatology*, Second Edition. Heidelberg: New York, Springer-Verlag.
128. LARIN, I. V. (1962). *Pasture Economy and Meadow Cultivation*, Publ. for Nat. Sci. Foundation, Washington, D.C., and Dept. Agric. by Israel Program for Scientific Translations. Publ. in Russian, Moscow/Leningrad, 1956. Publ. in English, Jerusalem, 641 pp.
129. LAZARIDES, M. (1960). Pastures of the North Kimberley Area, W.A. In Lands and Pastoral Resources of the North Kimberley Area, W.A., *Land Research Series*, No. 4: CSIRO, Australia, 109 pp.
130. ——, NORMAN, M. J. T. and PERRY, R. A. (1965). Wet-season development pattern of some native grasses at Katherine, N.T., *Tech. Paper 26, Div. Land Res. Reg. Survey*, CSIRO, Australia, pp. 19.
131. LEBRUN, J. (1947). *La végétation de la plaine alluviale au Sud du Lac Edouard.* Inst. Parcs. Nat. Congo Belg., 800 pp. (In 2 vols. 1–467, 469–800).
132. LEDGER, H. P. (1964). The role of wildlife in African agriculture, *E. Afr. Agric. For. J.*, 30: 137–41.

133. LEGRIS, P. (1960). Détermination des limites orientales du climat méditerranéen. Not published.
134. —— (1963). La végétation de l'Inde. Ecologie et Flore, *Tr. Sect. Sci. Techn. Inst. fr. Pondichéry*, VI: 596 pp.
135. —— and VIART, M. (1961). Bioclimates of South-India and Ceylon, *Tr. Sect. Sci. Techn. Inst. fr. Pondichéry*, III: 165–78.
136. MAHADEVAN, P. (1968). *Relations between climatic factors and animal production.* UNESCO Symposium on Methods in Agroclimatology, Reading, 23rd–30th July 1966. In Agroclimatological Methods. *Natural Resources Research Series* VII. UNESCO, Paris.
137. MAIRE, R. and MONOD, TH. (1950). Etudes sur la flore et la végétation du Tibesti, *Mémoire IFAN No.* 8: 141 pp.
138. —— and VOLKONSKY, N. (1945). Le passage du Sahara central au Sahara méridional (zone sahélo-saharienne) entre l'Adrar des Ifoghas et l'Aïr. *Trav. Inst. Rech. Sahar.*, III, 131–5.
139. MAYR, E. (1944). Wallace's line in the light of recent zoogeographic studies, *Quart. Rev. Biol.*, 19: 1–14.
140. MCKERRAL, A. (1937). The commoner grasses of Burma with notes on their agricultural importance and distribution, *Bull. Dept. Agric. Burma*, 20: 23 pp.
141. MCWILLIAM, J. R. (1964). Cytogenetics. In C. Barnard, *Grasses and Grasslands*, pp. 154–67.
142. MEHER-HOMJI, V. M. (1962). The bioclimates of India in relation to the vegetational criteria, *Bull. Bot. Surv. India*, 4: 105–12.
143. —— (1963). Les bioclimats du Sub-Continent indien et leurs types analogues dans le monde, *Tr. Sect. Sci. Techn. Inst. fr. Pondichéry*, 7: 254 pp.
144. —— (1964). Drought: Its ecological definition and phytogeographic significance. I. Ecological definition, *Trop. Ecol.*, 5: 17–31.
145. MEIGS, P. (1953). World distribution of arid and semi-arid homo-climates. In *Arid Zone Hydrology*, Paris, UNESCO.
146. MELVILLE, R. (1966). Continental drift, Mesozoic continents and the migrations of the Angiosperms, *Nature*, 211: 116–20.
147. MERRILL, E. D. (1936). Malaysian Phytogeography in relation to the Polynesian Flora. In Goodspeed, T. H. (ed.), *Essays in Geobotany in honour of William Albert Satchell*, 247–61.
148. MILNE-REDHEAD, E. (1954). Distributional ranges of flowering plants in tropical Africa, *Proc. Linn. Soc. Lond.*, 165: 25–35.
149. MONOD, TH. (1938). Notes botaniques sur le Sahara occidental et ses confins sahéliens, *La vie dans la région désertique nord tropicale de l'Ancien Monde.* Mémoire de la Société Biogéographique, 6: 351–74.
150. —— (1957). Les grandes divisions chorologiques de l'Afrique, *C.S.A./C.C.T.A. publication*, 24: London.
151. MOREAU, R. E. (1952). Africa since the Mesozoic, with particular reference to certain biological problems, *Proc. Zool. Soc. Lond.*, 121: 869–913.
152. —— (1963). Vicissitudes of the African biomes in the Late Pleistocene, *Proc. Zool. Soc. Lond.*, 141: 395–421.
153. MORTON, J. K. (1962). The upland floras of West Africa—their composition, distribution and significance in relation to climatic changes, *Comptes Rendus A.E.T.F.A.T. 1960*, 391–410.
154. MURAT, M. (1937). Végétation de la zone prédésertique en Afrique centrale, *Bull. Soc. Hist. Nat. Afr. Nord.*, 28: 19–83.
155. —— (1944). Esquisse phytogéographique du Sahara occidental, *Mémoires de l'Office national Anti-Acridien*, 1: 33 pp.

156. NAVEH, Z. (1966). Range research and development in the dry tropics with special reference to East Africa, *Herbage Abstracts*, 36: 77–85.

157. NEUBAUER, H. F. (1954/55). Versuch einer Kennzeichnung der Vegetationsverhältnisse Afghanistan, *Ann. Naturhist. Mus. Wien*, 60: 77–113.

158. NUMATA, M. (1965). *Ecological study and mountaineering of Mt. Numbur in Eastern Nepal, 1963*. Himalayan Expedition of Chiba University, Chiba, Japan, 166 pp. (Japanese).

159. NUTTONSON, M. Y. (1963). The Physical Environment and Agriculture of Central and South China, Hong Kong and Taiwan (Formosa), *Amer. Inst. Crop Ecol.*, 402 pp.

160. OKIY, C. E. O. (1960). Indigenous Nigerian food plants, *J. W. Afr. Sci. Assn.*, 6: 117–21.

161. PABOT, H. (1967). Report to Government of Iran on Pasture Development and Range Improvement through Botanical and Ecological Studies. FAO Report No. TA 2311. Mimeo 129 pp.

162. PAGOT, J. (1951–2). Production laitière en zone tropicale. Faits d'expérience en A.O.F., *Rev. Elev. Méd. vét. Pays trop.*, 5: 173–90.

163. —— (1964). Personal communication.

164. PARSA, A. (1957). Contributions to the knowledge of the useful plants and raw materials of Iran, *Materiae veg.*, 2: 174–83.

165. PAULME, DENISE (1953). *Les Civilisations africaines*. Presses Univ. de France, 124 pp.

166. PEDELABORDE, P. (1958). *Les moussons*. Armand Colin, Paris, 208 pp.

167. PENMAN, H. L. (1963). *Vegetation and Hydrology. Tech. Communic. 53.* Commonwealth Bureau of Soils, Harpenden. Comm. Agric. Bureaux, Farnham Royal, 124 pp.

168. PERRIN DE BRICHAMBAUT, G. and WALLEN, C. C. (1962). Report on an FAO/UNESCO /WMO Interagency Project on Agroclimatology in Semi-Arid and Arid Zones of the Near East, FAO, Rome, mimeo.

169. PERRY, R. A. (1960). Pasture Lands of the Northern Territory, Australia. CSIRO, Australia, *Land Research Series*, No. 5: 55 pp.

170. PHILLIPS, E. D. (1965). *The Royal Hordes: Nomad Peoples of the Steppes*. Thames and Hudson, London, 144 pp.

171. PHILLIPS, J. (1959). *Agriculture and Ecology in Africa*. Faber and Faber, London, 412 pp.

172. —— (1961). *The Development of Agriculture and Forestry in the Tropics*. Faber and Faber, London, 212 pp.

173. —— (1964). Shifting Cultivation—paper to 9th Technical Meeting of International Union for Conservation of Nature, Nairobi, September 1963. *Publications new series*, No. 4.

174. PICHI-SERMOLLI, R. (1955). Ethiopia/Somaliland/Kenya/Tanganyika. UNESCO Arid Zone Research Series VI: *Plant Ecology*, 302–60.

175. POLUNIN, N. (1960). *Introduction to Plant Geography and Some Related Sciences*. Longmans, Green and Co. Ltd. London. 640 pp..

176. POPOV, G. and ZELLER, W. (1963). Ecological Survey: Report on the 1962 Survey in the Arabian Peninsula, *FAO Progress Report*, No. UNSF/DL/ES/6 Rome, 99 pp.

177. PORTÈRES, R. (1951). Géographie alimentaire, berceaux agricoles et migrations des plantes cultivées en Afrique intertropicale, *C. R. Sess. Biogeogr.*, 239–40, 16–21.

178. QUADRI, S. M. A. (1955). Ecological study of the riverain forests of Sind, *Pak. J. For.*, 5: 241–9.

179. QUEZEL, P. (1958). Mission botanique au Tibesti, *Mémoire no. 4 de l'Inst. de Rech. sahariennes.*

180. RAHEJA, P. C. (1965). Influence of climatic changes on vegetation of the arid zone in India, *Ann. Arid Zone*, 4: 64–73.

181. RAMSAY, J. M. and ROSE INNES, R. (1963). Some quantitative observations on the effects of fire on the Guinea savannah vegetation of northern Ghana over a period of eleven years, *African Soils*, 8: 41–85.

182. RANGANATHAN, C. R. (1938). Studies in the ecology of the shola grassland vegetation of the Nilgiri plateau, *Indian For.*, 64: 523–41.

183. RATTRAY, J. M. (1960). *The Grass Cover of Africa, FAO Agric. Study* No. 49, Rome (Eng. Fr. Sp.), 168 pp. with accompanying map.

184. RAUNKIAER, C. (ed. TANSLEY, A. G.) (1934). *The Life Forms of Plants and Statistical Plant Geography*, Oxford, xvi + 632.

185. RECHINGER, K. H. (1963). *Flora of Lowland Iraq*. Cramer Verlag, Weinheim an der Bergstrasse.

186. RHIND, D. (1945). *The Grasses of Burma.* Baptist Mission Press, Calcutta, 99 pp.

187. RISOPOULOS, S. (1964). *Project for development of Okpara Farm as: a breeding ranch; a centre for finishing local cattle; a centre for pig production.* FAO, PL/1/64, 41 pp., mimeogr.

188. ROBBINS, R. G. (1960). The vegetation of New Guinea, *Australian Territories* 1, no. 6, pp. 1–12.

189. —— (1960). The anthropogenic grasslands of Papua and New Guinea, *Proc. UNESCO Symp. on Impact of Man on Humid Tropics Vegetation, New Guinea, 1960*, 313–29.

190. —— and PULLEN, R. (1965). Vegetation of the Wabag-Tari Area. In Lands of the Wabag-Tari Area, Papua–New Guinea, *Land Research Series*, 10, CSIRO, Australia, 142 pp. with map.

191. ROSSETTI, C. (1965). Ecological Survey Mission to West Africa. Studies on the Vegetation (1959 and 1961). Discussions and conclusions. FAO, Rome, *UNSF/DL/ES/5*, mimeogr., 77 pp.

192. RUSSELL, E. W. (ed.) (1962). *The natural resources of East Africa.* Nairobi: D. A. Hawkins in association with East African Literature Bureau, 144 pp. + maps.

193. —— (1968). *Climate and Soils.* Paper to UNESCO Symposium on Methods in Agroclimatology, Reading, 23rd–30th July 1966. In Agroclimatological Methods. *Natural Resources Research Series* VII, UNESCO, Paris.

194. SACCO, T. (1964). Contribution to the study of the flora of pastures in some zones of Baluchistan and of Iranian Sistan, *Allionia*, 10: 53–87.

195. SANDFORD, R. H. D. (1964). Report to Government of Ethiopia on Livestock Production. *FAO EPTA Report* No. 1925, mimeogr., FAO, Rome, 26 pp.

196. SATYANARAYAN, Y. (1963). Ecology of the Central Luni Basin, Rajasthan, *Annals of Arid Zone*, 2: 82–97.

197. —— and GAUR, Y. D. (1965). Sociological variations in floristic composition of the vegetation in the arid zone. Personal communication.

198. SCHNELL, R. (1957). *Plantes alimentaires et Vie agricole de l'Afrique noire.* Ed. Larose, Paris, 222 pp.

199. SELOD, Y. I. (1961). *Bioclimats et Végétation du Pakistan occidental.* Thèses présentées à la Faculté des Sciences de l'Université de Toulouse, 116 pp.

200. SETH, S. K. (1963). In Change of climate, with special reference to the arid zones. ARID ZONE RESEARCH SERIES XX, UNESCO, Paris, pp. 488.

201. —— and KHAN, M. A. WAHEED (1959). Bioclimate and plant introduction in dry zone, *Ind. For.*, 85: 376–84.

202. SHANKARNARAYAN, K. A. (1962–3). *Systematic Botany. Central Arid Zone Research Inst. Sci. Progress Report*, 56–8.
203. SLATYER, R. O. (1960). Agricultural Climatology of the Katherine Area, N.T., *Tech. Pap. 13, Div. Land Res. Reg. Surv.* CSIRO, Australia, 39 pp.
204. —— (1968). *The use of soil water balance relationships in agroclimatology.* Paper to UNESCO Symposium on Methods in Agroclimatology, Reading, 23rd–30th July 1966. In Agroclimatological Methods. *Natural Resources Research Series* VII. UNESCO, Paris.
205. SMITH, J. (1949). Distribution of tree species in the Sudan in relation to rainfall and soil texture. *Sudan Gov. Agr. Pub. Bull.*, 4, Khartoum.
206. SPECHT, R. L. (1958). The Gymnospermae and Angiospermae collected on the Arnhem Land Expedition. In Specht and Mountford (eds.), *Records of the American–Australian Scientific Expedition to Arnhem Land*, Vol. 3. *Botany and Plant Ecology*. Melbourne Uni. Press, 185–317.
207. —— (1958). The climate, geology, soils and plant ecology of the northern portion of Arnhem Land. In Specht and Mountford (eds.), *Records of the American–Australian Scientific Expedition to Arnhem Land*, Vol. 3: *Botany and Plant Ecology*. Melbourne Uni. Press, 333–414.
208. —— (1958). The geographical relationships of the flora of Arnhem Land. In Specht and Mountford (eds.), *Records of the American–Australian Scientific Expedition to Arnhem Land*, Vol. 3: *Botany and Plant Ecology*. Melbourne Uni. Press, 415–78.
209. —— and MOUNTFORD, C. P. (1958) (eds.). *Records of the American–Australian Scientific Expedition to Arnhem Land*, Vol. 3: *Botany and Plant Ecology*. Melbourne Uni. Press, Carlton, Victoria, Australia, xv + 522 pp.
210. SPECK, N. H. and LAZARIDES, M. (1964). Vegetation and pastures of the West Kimberley Area. In General Report on Lands of the West Kimberley Area, W.A., *Land Research Series*, No. 9. CSIRO, Australia, 219 pp.
211. STAMP, L. D. (1925). The Vegetation of Burma from an Ecological Standpoint, *Univ. Rangoon Res. Monogr.*, No. 1, 58 pp.
212. —— (1962). *Asia: A Regional and Economic Geography*, (eleventh edition). Methuen and Co. Ltd., London, 730 pp.
213. STEWART, G. and HUTCHINGS, S. S. (1936). The point-observation-plot (sq. foot density) method of vegetation survey, *J. Amer. Soc. Agron.*, 28: 714–26.
214. SUBRAHMANYAM, V. P. (1956). Climatic types of India according to the rational classification of Thornthwaite, *Indian J. Met. Geophys.*, 7: 253–64.
215. SYMINGTON, C. F. (1933). The study of secondary growth on rain forest sites in Malaya, *Malay. For.*, 2: 107–17.
216. TALBOT, LEO M., LEDGER, H. P. and PAYNE, W. J. A. (1961). The possibility of using wild animals for animal production in the semi-arid tropics of East Africa, *Festschrift zum VIII Intern. Tierzuchtkongress in Hamburg*, 205–10.
217. TAYLOR, B. W. (1964). Vegetation of the Buna–Kokoda Area. In Lands of the Buna–Kokoda Area, Territory of Papua and New Guinea, *Land Research Series*, No. 10. CSIRO, Australia, 113 pp. with map.
218. —— (1964). Vegetation of the Wanigela–Cape Vogel Area. In Lands of the Wanigela–Cape Vogel Area, Papua–New Guinea, *Land Research Series*, No. 12. CSIRO, Australia, 99 pp. with map.
219. THORNTHWAITE, C. W. (1943). The climates of the earth, *Geogr. Rev.*, 33: 433–40.
220. —— (1948). An approach toward a rational classification of climate, *Geog. Rev.*, 38: 55–94.
221. TREGEAR, T. R. (1965). *A Geography of China*. Uni. London Press, 342 pp.

222. TREWARTHA, G. T. (1963). *The Earth's Problem Climates*. Methuen, London. 334 pp.

223. TSING LIU (1963). Ecological study on the high mountain meadow of Mt. Hsiao-Hsüeh, *Bull. Taiwan Forestry Res. Inst.*, No. 92: 1–16.

224. U.S.S.R. MINISTRY OF GEOLOGY AND CONSERVATION OF NATURAL RESOURCES (1960). Agricultural Atlas of U.S.S.R., Moscow.

225. UNITED NATIONS EDUCATIONAL, SCIENTIFIC AND CULTURAL ORGANIZATION SCIENCE COOPERATION OFFICE FOR SOUTH EAST ASIA and COUNCIL FOR SCIENCES OF INDONESIA (1958). Proceedings of Symposium on Humid Tropics Vegetation, Indonesia, December 1958, 312 pp. (Caxton Press, New Delhi).

226. —— and GOVERNMENT OF SARAWAK (1965). Symposium on Ecological Research in Humid Tropics Vegetation. Kuching, Sarawak, July 1963. Tokyo Press Co., Japan, 376 pp.

227. UVAROV, B. (1966). *Grasshoppers and Locusts*, Vol. 1. Cambridge Uni. Press, 492 pp.

228. VAN MEEUWEN, M. S., NOOTEBOOM, H. P. and VAN STEENIS, C. G. G. J. (1961). Preliminary revisions of some genera of Malaysian Papilionaceae I, *Reinwardtia*, 5: 419–56.

229. VAN STEENIS, C.G.G.J. (1933). On the origin of the Malaysian mountain flora. I. Facts and statement of the problem. *Bull. Jard. Bot. Btzg.* (III) 13, 135-262.

230. —— (1934). On the origin of the Malaysian mountain flora. II. Altitudinal zones, general considerations and renewed statement of the problem. *Bull. Jard. Bot. Btzg.* (III) 13, 289-417.

231. —— (1936). On the origin of the Malaysian mountain flora. III. Analysis of floristic relationships, pt. i. The Sumatran track. *Bull. Jard. Bot. Btzg.* (III) 14, 56-72.

232. —— (1958). *Vegetation Map of Malaysia.* UNESCO, Paris, Humid Tropics Project.

233. VASILJEV, V. N. (1958). The origin of the flora and vegetation of the Far East and Eastern Siberia. In *Materials on the History of the Flora and Vegetation of the U.S.S.R.*

234. VESEY-FITZGERALD, D. F. (1955). Vegetation of the Red Sea coast south of Jedda, Saudi Arabia, *J. Ecol.*, 43: 477–89.

235. —— (1957). The vegetation of the sea coast north of Jedda, Saudi Arabia, *J. Ecol.*, 45: 547–62.

236. —— (1957). The vegetation of central and eastern Arabia, *J. Ecol.*, 45: 779–98.

237. —— (1963). Central African grasslands, *J. Ecol.*, 51: 243–74.

238. —— (1965). The utilization of natural pastures by wild animals in the Rukwa Valley, Tanganyika, *E. Afr. Wildl. J.*, 3: 38–48.

239. VOROSHILOV, V. N. (1966). *Flora of the Soviet Far East.*

240. WALLEN, C. C. (1968). *Agroclimatological Studies in the Levant.* Paper to UNESCO Symposium on Methods in Agroclimatology, Reading, 23rd–30th July 1966. In Agroclimatological Methods. *Natural Resources Research Series* VII, UNESCO, Paris.

241. WANG, CHI-WU (1961). *The Forests of China.* Maria Moors Cabot Foundation Publication Series, No. 5. Harvard, Mass., 313 pp.

242. WARBURG (1891). Beiträge zur Kenntnis der papuanischen Flora. *Bot. Jahrb.*, 13: 230–455.

243. WARD, F. K. (1936). A sketch of the vegetation and geography of Tibet, *Proc. Linn. Soc. London*, 148: 133–60.

244. —— (1948). Tibet as a grazing land, *Geogr. J.*, 110:60–75.
245. WEAVER, J. E. (1961). *Return of Midwestern grassland to its former composition and stabilization.* (Occ. papers, C. C. Adams Center for Ecol. Studies, No. 3), 15 pp.
246. —— and ALBERTSON, F. (1956). *Grasslands of the Great Plains. Their Nature and Uses.* Lincoln (Nebraska), Johnsen Publishing Co., 395 pp.
247. WHITE, F. (1965). The savannah woodlands of the Zambezi and Sudanian domains: an ecological and phytogeographical comparison, *Webbia*, 19: 651–81.
248. WHYTE, R. O. (1955). Report to the Government of Japan on Pasture Development. FAO ETAP Report No. 462. Mimeo. 42 pp.
249. —— (1961) Evolution of land use in south-western Asia. In A History of Land Use in Arid Regions, *Arid Zone Research Series* XVII, UNESCO, Paris, 388 pp. (58–118).
250. —— (1963). Report of the Government of Japan on Pasture and Fodder Development. FAO EPTA Report No. 1822. Mimeo. 41 pp.
251. —— (1963). The significance of climatic change for natural vegetation and agriculture. In Change of climate with special reference to the arid zones: Proceedings of the UNESCO/WMO Rome Symposium, *Arid Zone Research Series* XX. UNESCO, Paris, pp. 381–6.
252. —— (1964). *The Grassland and Fodder Resources of India.* Indian Council of Agricultural Research Sci. Mon. 22 (revised edition), 553 pp.
253. —— (1966). *The Use of Arid and Semi-Arid Land.* In *Arid Lands*, chap. 15. Methuen, London/UNESCO, Paris, 301–61.
254. ——, DABADGHAO, P. M. and SHANKARNARAYAN, K. A. (1969). *The Grass Cover of India.* Indian Council of Agricultural Research Sci. Mon. (in press).
255. WILD, H. (1956). The principal phytogeographic elements of the Southern Rhodesian flora, *Proc. Trans. Rhodesia Sci. Ass.*, 44: 1–12.
256. —— (1964). The endemic species of the Chimanimani Mountains and their significance, *Kirkia*, 4: 125–64.
257. —— (1965). Additional evidence for the Africa–Madagascar–India–Ceylon land-bridge theory with special reference to the genera *Anisopappus* and *Commiphora*, *Webbia*, 19: 497–505.
258. WILLIAMS, R. J. (1955). Vegetation Regions. In *Atlas of Australian Resources* (map and explanatory notes).
259. WOOD, J. G. (1959). The Phytogeography of Australia. In A. Keast, R. L. Crocker and C. S. Christian (Eds). *Biogeography and Ecology in Australia.* Uitgeverij Dr. W. Junk. The Hague. 640 pp. (291–302).
260. —— and WILLIAMS, R. J. (1960). Vegetation of Australia, chapter VI In *The Australian Environment*, 3rd ed. Cambridge Uni. Press, London and New York, 151 pp. (66–84).
261. ZINDERER BAKKER, E. M. VAN (1964). Chapter 10, Phytogeography, of *Palynology in Africa.* Eighth Report: 54–62.
262. ZOHARY, M. (1935). Die phytogeographische Gliederung der Flora der Halbinsel Sinai II. Spezieller Teil. 1. Die Elemente. Das Saharo-sindische Element, *Beih. Bot. Centralbl.*, 52 B., 596.
263. —— (1963). On the geobotanical structure of Iran, *Bull. Res. Counc. Israel* (*D*). Suppl. II, 113 pp.
264. ZOLOTAREVSKY, B. and MURAT, M. (1938). Divisions naturelles du Sahara et sa limite méridionale, *La vie dans la région désertique de l'ancien monde.* Mémoire de la Société de Biogéographie, 6: 335–50.

265. Chang, Jen-Hu (1967). The Indian summer monsoon, *Geogr. Rev.*, 57: 373-96
266. Bor, N. L. and Melderis, A. (1965). Gramineae. In *Symbolae Afghanicae* Vol. VI, Biologiske Sknfter Vol. 14: 4. Royal Danish Academy of Sciences and Letters, Copenhagen, 36–94.
267. Cocheme, J. and Franquin, P. (1967). *An Agroclimatology Survey of a Semiarid Area in Africa south of the Sahara.* WMO Technical Note No. 86, Geneva, 136 pp.
268 Audry, P. and Rossetti, Ch. (1962). Observations sur les sols et la végétation en Mauritanie du SE et sur le bordure adjacente du Mali. FAO, Rome.
269. Gaussen, H., Legris, P. and Blasco, F. (1967). Bioclimats du Sud-Est Asiatique. *Tr. Sect. Sci. Techn. Inst. fr. Pondichéry*, III, 114 pp.
270. Whyte, R. O. (1966). Géographie des graminées, *Agron. Trop.* 6/7, 778–85.
271. —— (1968). *Land, Livestock and Human Nutrition in India.* Special Study Frederick A. Praeger. New York and Washington.

INDEX OF PLANT NAMES

For the centre of the transect, it has been possible to check names of species of the Gramineae with N. L. Bor's *Grasses of Burma, Ceylon, India and Pakistan*. For other plant families and for the western and eastern parts of the transect, the names used by the many authors cited have been adopted.

INDEX OF PLANT NAMES

INDEX OF GEOGRAPHICAL NAMES AND SUBJECTS